OECD Studies on Water

Water Governance in Cities

OECD
BETTER POLICIES FOR BETTER LIVES

Please cite this publication as:
OECD (2016), *Water Governance in Cities*, OECD Studies on Water, OECD Publishing, Paris.
http://dx.doi.org/10.1787/9789264251090-en

ISBN 978-92-64-25108-3 (print)
ISBN 978-92-64-25109-0 (PDF)

Series: OECD Studies on Water
ISSN 2224-5073 (print)
ISSN 2224-5081 (online)

Co-published by IWA Publishing
Alliance House, 12 Caxton Street, London SW1H 0QS, UK
Tel: +44 (0)20 7654 5500, Fax: +44 (0)20 7654 5555
publications@iwap.co.uk
www.iwapublishing.com

ISBN: 9781780408248 (Paperback)
ISBN: 9781780408255 (eBook)

The statistical data for Israel are supplied by and under the responsibility of the relevant Israeli authorities. The use of such data by the OECD is without prejudice to the status of the Golan Heights, East Jerusalem and Israeli settlements in the West Bank under the terms of international law.

Photo credits: Cover © Taro Yamada/Corbis; © Lindwa/Shutterstock.com.

Corrigenda to OECD publications may be found on line at: *www.oecd.org/about/publishing/corrigenda.htm*.

Foreword

Many cities around the world are increasingly subject to water-related risks. Copenhagen and New York City were hit by floods in 2011 and 2012 with severe economic losses. Mexico City is challenged by serious aquifer contamination, while cities in California are in the midst of the worst drought in the state's history. These crises are only a harbinger of things to come. Future crises will be exacerbated by economic, population, climate and urbanisation trends. Technical solutions to manage supply and demand exist and are generally well-known. Today, the key challenge is to align incentives and choose the relevant policy instruments to move from crisis response to adequate management and anticipation of water risks. This implies a critical role for robust public policies across levels of government and for shared responsibility among stakeholders as specified in the OECD *Principles on Water Governance*.

The recently adopted Sustainable Development Goals (SDGs) set milestones for the years to come and call for action in relation to water management in cities: *ensuring availability and sustainable management of water and sanitation for all* (SDG 6); *making cities and human settlements inclusive, safe, resilient and sustainable* (SDG 11). They also call for establishing or strengthening *inclusive societies* (SDG 16) and *global partnerships for sustainable development* (SDG 17). A prominent feature of the SDGs is that they are global, and as such apply to both developed and developing economies. These goals are very much in line with OECD's main message on water over the past decade: current levels of service delivery and water security should not be taken for granted and simultaneous action is needed on three fronts: infrastructure, institutions and information.

The present report, *Water Governance in Cities,* primarily targets local decision makers and the crucial role they play given that water is mostly managed at the local level and that 50% of the world's population lives in cities. Since 2009, the OECD Water Governance Programme has been assessing the capacity of governance systems to handle current and future water challenges. Policy guidance was provided to national governments through benchmarks across 17 OECD countries (2011) and 13 Latin American countries (2012), as well as policy dialogues with Mexico (2013), the Netherlands (2014), Jordan (2014), Tunisia (2014) and Brazil (2015).

Water Governance in Cities builds on a survey of 48 cities across OECD and non-OECD countries, the preliminary results of which were published in the report *Water and Cities: Ensuring Sustainable Futures* developed as part of the OECD Horizontal Water Programme. It also draws extensively on the OECD's work on urban development and territorial indicators to offer a clustering of cities by size, spatial organisation, demographic dynamics and metropolitan governance.

Different cities face specific water challenges and have varying capacities to respond. And yet, for cities having similar characteristics, place-based responses can be developed, and peer-to-peer dialogue can help progress. This report proposes a "3Ps" co-ordination framework around *policies, people and places* to suggest responses to the challenges identified. It showcases best practices to promote a strategic vision across sectors, to engage with stakeholders and to foster integrated urban water management in cities and their hinterlands, through rural-urban partnerships and metropolitan governance.

Angel Gurría
OECD Secretary-General

Acknowledgements

This report is the result of a two-year project spearheaded and managed by Aziza Akhmouch, Head of the Water Governance Programme, together with Oriana Romano, Policy Analyst, within the Regional Development Policy Division, led by Joaquim Oliveira-Martins, at the Public Governance and Territorial Development Directorate directed by Rolf Alter.

The authors are particularly grateful to the 48 cities surveyed (see the list in Annex A) for their responses to the questionnaire and feedback throughout the project. Elizabeth Sosa Altamirano, Secretary to the Permanent Representative of Mexico to the OECD, and Jinjing Chen, OECD Global Relations Secretariat, were instrumental in engaging and liaising with cities from Mexico and China. The report benefited from valuable comments from several OECD colleagues, in particular Luiz De Mello, Joaquim Oliveira Martins, William Tompson, Karen Maguire, Xavier Leflaive, Céline Kauffmann, Guillaume Gruère, Monica Brezzi, Rudiger Ahrend and Paolo Veneri. Guidance and suggestions were also provided on the design of the questionnaire and data processing by Claire Charbit, Daniel Sanchez Serra, Eric Gonnard, Isabelle Chatry, Soo-Jin Kim and Abel Schumann. Delphine Clavreul, Antonio Canamas Catala, Shahnaaz Parker, Yuan Chang and Christoph Brodnik from the OECD Water Governance Programme contributed with comments, best practice collection and fact-checking.

Preliminary findings were presented on several occasions, in particular at the European Water Association 10th Conference on Water in Cities (Brussels, 18 November 2014); the 7th World Water Forum sessions on "Living with Water in Cities of Tomorrow" and "Boosting Sustainable Local Water Management through Multi-level Governance and Financing" (Daegu and Gyeongbuk, Korea 12-17 April 2015); the Water Conference 2015 (Athens, 16 June 2015); the 3rd Global Water Operators Partnership Congress (Barcelona, 17 September 2015); and at the Mediterranean Economic Week (Marseille, 5 November 2015).

Advanced results were discussed during an OECD webinar with surveyed cities (26 October 2015) and at the 6th meeting of the OECD Water Governance Initiative (2 November 2015). Written comments were provided by Susana Neto, Technical University of Lisbon; Salvatore Ercolano and Giuseppe Lucio Gaeta, University of Naples "L'Orientale"; Peter Gammeltoft, former Head of Unit for Water at the European Commission; Milo Fiasconaro, Aqua Publica Europea; Gari Villa-Landa Sokolova, Spanish Association of Water Utilities; Bernard Barraqué, Centre national de la Recherche Scientifique; Wang Jianping, Ministry of Water Resources, China; Antonio Martinez Nieto, Murcia Water Agency; Dominique Gatel, Veolia; Thomas Hartmann, Utrecht University; and Donal O'Leary, Transparency International, and Uta Wehn, UNESCO Institute for Water Education.

The draft report was peer-reviewed at the 34th meeting of the OECD Regional Development Policy Committee (4 November 2015). Special thanks are herein conveyed to Rick Roelofs, Ministry of Economic Affairs of the Netherlands, Birgit Wilder and Marlies Stubits, Permanent Representative of Austria to the OECD, and Keith Thorpe, Department for Communities and Local Governments of the United Kingdom, for their comments on that occasion.

The report was edited by Hilary Delage. Ulrike Chaplar, Julie Harris and Pilar Philip prepared the manuscript for publication.

Table of contents

Tables

Figures

Boxes

Acronyms and abbreviations

AEEGSI	National Authority of Electricity, Gas and Water Services (Italy) (*Autorità per l'energia elettrica, il gas ed il sistema idrico*)
AMB	Metropolitan Area of Barcelona (Spain) (*Àrea Metropolitana Barcelona*)
ARPA	Regional Agency for the Protection of the Environment (Italy) (*Agenzia regionale per la prevenzione e l'ambiente*)
ATERSIR	Territorial Agency of Emilia-Romagna for Water Services and Waste (Italy) *(Agenzia territoriale dell'Emilia-Romagna per i servizi idrici e rifiuti)*
ATO	Optimal territorial area (Italy) (*Ambito Territoriale Ottimale*)
BLUE AP	Bologna Local Urban Environment Adaptation Plan for a Resilient City (Italy)
BoD	Board of Directors
BOD5	Biochemical Oxygen Demand
CEA	State Water Commission (Mexico) *(Comisión Estatal de Agua: CEA Querétaro)*
CMM	Montreal Metropolitan Community (Canada) *(Communauté Métropolitaine de Montréal)*
CODEME	Council for the Metropolitan Development of the City of Acapulco (Mexico) (*Consejo para el Desarrollo Metropolitano del Municipio de Acapulco*)
DEP	Department of Environmental Protection (United States)
EPAL	Water supply in the Greater Lisbon area (Portugal) (*Empresa Portuguesa das Águas Livres*)
ERSAR	Water and Waste Services Regulation Authority (Portugal) (*Entidade Reguladora dos Serviços de Águas e Resíduos*)
EPA	Environmental Protection Agency (United States)
EU	European Union
EYATH	Thessaloniki Water Supply and Sewerage Company (Greece) *(Εταιρεία Ύδρευσης και Αποχέτευσης Θεσσαλονίκης Α.Ε.)*
EYDAP	Athens Water Supply and Sewerage Company (Greece) *(Εταιρεία Ύδρευσης και Αποχέτευσης Πρωτευούσης)*
FNAB	National Federation of Organic Agriculture (France) *(Fédération Nationale d'Agriculture Biologique des régions de France)*
FUAs	Functional Urban Areas
GCVSDPA	Glasgow and the Clyde Valley Strategic Development Planning Authority (United Kingdom)
GIS	Geographic Information System
GDP	Gross Domestic Product
HERA	Water Service Operator Bologna (Italy) *(Holding Energia Risorse Ambiente)*

HOFOR	Greater Copenhagen Utility (Denmark) *(Hovedstadsområdets Forsyningsselskab)*
HRADF	Hellenic Republic Asset Development Fund (Greece)
HRWS	Human Rights to Water and Sanitation
IBNET	International Benchmarking Network for Water and Sanitation Utilities
INTERAPAS	Metropolitan Inter-municipal Organism of Water, Sewage and Sanitation of San Luis Potosi (Mexico) (*Organismo Intermunicipal Metropolitano de Agua Potable, Alcantarillado, Saneamiento y Servicios*)
ICTs	Information and Communications Technologies
IWA	International Water Association
NGO	Non-Governmental Organisation
NYC	New York City
OECD	Organisation for Economic Co-operation and Development
OFWAT	The economic regulator of the water sector in England and Wales
ÖWAV	Austrian Water and Waste management Association (Austria) (*Österreichischer Wasser- und Abfallwirtschaftsverband*)
ÖVGW	Austrian Association for Gas and Water (Austria) (*Österreichische Vereinigung für das Gas- und Wasserfach*)
PPPs	Public-Private Partnerships
PPP	Purchasing Power Parity
PUB	Public Utilities Board
RBOs	River Basin Organisations
RWBT	Rondout-West Branch Tunnel
SABESP	São Paulo Water and Waste Management Company (Brazil) (*Companhia de Saneamento Básico do Estado de São Paulo*)
SDG	Sustainable Development Goal
TRUST	Transitions to the Urban Water Services of Tomorrow
TWM	Total Water Management
TWRP	Tuas Water Reclamation Plant
UU	United Utilities Liverpool (United Kingdom)
Vewin	Association of Dutch water companies (The Netherlands) *(Vereniging van drinkwaterbedrijven in Nederland)*
WB	World Bank
WHO	World Health Organization
WSS	Water Supply and Sanitation

Executive summary

Too much, too little or too polluted: more and more, this characterises the key water challenges facing cities. Urban areas currently host about 50% of the global population, projected to reach over 60% by 2050. Over this same period, water demand will increase by 55% globally, and about 4 billion people will be living in water-stressed areas. This means that fierce competition across different categories of water users – particularly agriculture, energy and urban dwellers – is unavoidable. It also means that if nothing changes, water security will be increasingly threatened.

Because water as the livelihood for current and future generations is of public interest, there is a critical role for public policies to address pressing and emerging challenges in this area. Good urban water governance, in particular, is essential for managing water-related risks in a timely manner and at an acceptable cost, so that the next generation does not inherit liabilities and costs. Doing this right requires assessing the range of political, institutional and administrative rules, practices and processes (formal and informal) through which decisions are taken and implemented, consulting stakeholders, as well as holding decision-makers accountable for water management. Understanding the current state of play of urban water governance and adjusting where necessary is important. Water crises are often governance crises, involving more than hydrology, finance and infrastructure issues. They frequently reveal deficits in terms of who does what, at what level of government, and how and why public policies are designed and implemented.

This report provides qualitative and quantitative evidence on the performance of current water governance systems in cities, as well as best practices that can inspire other cities. Chapter 1 proposes an analytical framework to assess urban water governance and groups of cities facing similar types of challenges. Chapter 2 analyses the range of urban, climatic, demographic and economic factors that impact water management in cities. Chapter 3 maps who does what and at what level and draws on recent trends in the allocation of roles and responsibilities across governments and service providers. Chapter 4 diagnoses the main governance gaps in the effective design, implementation and evaluation of urban water policies. Chapter 5 suggests a framework and good practices for co-ordinating across multiple scales, authorities, and policy domains.

Building on a survey of 48 cities from OECD countries as well as emerging economies, the report shows that significant progress has been achieved in urban water management, but important challenges remain. For instance among cities surveyed, the average share of wastewater treated was 90% in 2012 compared to 82% in 1990; per capita domestic water consumption decreased by 20% between 2000 and 2012, while 98% of the population had access to drinking water in 2012, compared to 94% in 1990. However a number of challenges are noteworthy:

- Inequality in access to services and affordability remain important in some places. Unconnected fractions of population to water systems are reported in some urban and peri-urban areas of OECD countries (e.g. Greece, Italy, Mexico,

Portugal). The lowest access rates to sanitation services among surveyed cities are reported in Belo Horizonte (75%) and Veracruz (79%). Some targeted groups (e.g. poor people; those living in disadvantaged areas, ethnic minorities) are still in need of solutions to tackle affordability.

- Despite progress in water quality, 75% of surveyed cities identified water pollution as a challenge. Continued investments in wastewater treatment, together with improvements in irrigation systems are needed, especially as in some OECD countries, part of the population is still connected to a wastewater treatment plant with primary treatment only or to a sewerage network without treatment.
- Cities are increasingly exposed to water-related disasters. Recent episodes of floods that occurred in Copenhagen (2011) and New York (2012) generated enormous damages and economic losses. Projections show that nearly 20% of the world's population will be at risk from floods by 2050, while several cities are already suffering from the consequences of heavy droughts, even in water-rich countries such as Brazil.
- More than 90% of the cities surveyed reported ageing or lacking infrastructure, which threatens universal coverage of drinking water and sanitation and diminishes the capacity to protect citizens against water-related disasters. This is compounded by factors such as climate change, economic crisis, demographic growth and urban trends.
- Capacity is often the Achilles' heel of sub-national governments: 65% of surveyed cities reported the lack of staff and managerial competencies as a challenge. The effective implementation of water responsibilities is also threatened by unstable or insufficient revenues of local governments, whose budgets have been further tightened by the financial crisis while they continue to be mostly responsible for public investment in the OECD region.

Key messages

The high quality of urban water services in OECD countries is threatened by a massive investment backlog impeding the upgrading, renewal and maintenance of water-related infrastructure. There is a need to address public investment issues including multi-level co-ordination and capacity challenges; foster cross-sectoral approaches to infrastructure; adopt an approach that encompasses multiple purposes, especially water, agriculture, energy, environmental protection and spatial planning; manage trade-offs across water users in rural and urban areas and between current and future generations in terms of who pays for what; and reduce investment needs by ensuring stable regulatory frameworks to catalyse finance and enhance efficiency. A good practice is the EPA Water Infrastructure and Resiliency Finance Center set up in April 2015 to help US municipalities efficiently use federal and local funds for water infrastructure, explore financing options and showcase best practices. Another example is Singapore's plan to co-locate the Tuas Water Reclamation Plant and the Integrated Waste Management Facility by 2017 to allow an integrated management of water and solid waste, while maximising both energy and resources recovery.

Current levels of water security are jeopardised by climate change, urbanisation, and demographic and economic trends. To respond, it is important to raise awareness among citizens and policy makers to trigger policy and behavioural change; engage with stakeholders, including property developers and long-term institutional investors, to build consensus on the acceptable level of risk and secure willingness to pay for water services; and strengthen water-related data and information for more robust early-warning systems, monitoring and evaluation. Good practices include the Bologna Local Urban Environment Adaptation Plan for a Resilient City (BLUEAP), which involved 150 stakeholders, 70 project ideas and 6 pilot actions to come, amongst others, with solutions to water scarcity. The Water Observatory of the municipality of Paris provides a multi-stakeholder consultative platform prior to discussions at the City Council. Information and communications technologies are used to display water quality and quantity data in a number of cities including Marseille, while communication campaigns such as "Max 100" in Copenhagen raised awareness of citizens and fostered water savings.

Roles and responsibilities for managing water in cities are spread across different levels of government and a broad range of stakeholders such as public authorities, service providers, regulators, and river basin organisations. Even in highly decentralised contexts, national governments have a role to play in setting proper incentives and frameworks for urban water governance. Clarifying who does what and at what level of government can help identify potential mismatches, duplications or grey areas and assist in co-ordinating the actions of multiple players in an effective, efficient and inclusive way. Strategies for water governance at the metropolitan level may offer interesting models for co-ordination: the Metropolitan Glasgow Strategic Drainage Partnership gathers local authorities and national agencies to address the risks of flooding and improve water quality; the Metropolitan Authority of Barcelona has fostered an integrated perspective across local governments as well as shared infrastructure and expenses, while Nantes Métropole has been engaging stakeholders to improve local services.

Water in cities is affected by decisions taken in other sectors and vice versa, in particular agriculture, energy, finance, solid waste, transport and land use. There is a need to ensure that water is recognised as a key factor of sustainable growth in cities. Such a strategic vision is essential for strengthening policy coherence for an integrated urban water policy, mitigating split incentives whereby those generating future liabilities do not bear related costs, and fostering a whole-of-government approach that builds on horizontal and vertical co-ordination. For instance, in the Netherlands "water assessments" are carried out in municipalities to factor in water-related stakes and costs in spatial planning decisions; the city of Cologne co-ordinates water and spatial planning for new building areas to prevent flood damages because of heavy rainfalls; the city of Milan is combining water resources management and waste management under the umbrella of the project "Smart Water Resource Management"; Eau de Paris put in place concrete actions to promote organic agriculture for the preservation of water and natural resources. New York's "Green Infrastructure Program" promotes the resilience of the drainage system.

In most cases, hydrological boundaries cut across city administrative perimeters, requiring a functional approach to water management. Further efforts are needed to foster greater co-operation between cities and their hinterlands, and among cities of a given metropolitan area. The scale at which water is managed depends on the function: it can be at the local or metropolitan level for drinking water and sanitation; at the sub-basin or basin level for water resources management; or at higher levels for flood protection. Multiple scales thus need to be combined within integrated basin governance systems to foster efficient use of water resources, information sharing and cost savings. Good practices include contracts between the municipality of Paris, authorities in the hinterland

and farmers to foster co-operation between supplying areas in terms of water resources and the urban core; a multi-stakeholder committee in Montreal helped improve the quality of discharged water in catchment areas; in New York City, an agreement with watershed communities and other authorities helped to preserve both water quality and the economic dynamism of the area.

A "3Ps" framework can help respond to the above challenges:

- **Policy:** co-ordinate water across policies must favour inter-sectoral complementarities while efficiently allocating resources and building capacities.
- **People:** people's awareness of current and future water risks must be raised and inclusive decision-making enhanced.
- **Places:** place-based understanding is required to overcome territorial mismatches and favour co-operation between cities and their surroundings.

This framework seeks to foster multi-level interactions across the water policy cycle and beyond, while putting special emphasis on metropolitan governance, rural-urban partnerships and stakeholder engagement.

Chapter 1

Urban water governance today – Setting the scene

Intensifying water competition across users (e.g. households, farmers, urban dwellers and industry); renewing ageing infrastructure, restoring the ecological status of water bodies; preserving ecosystems; and maintaining adequate access to, and quality of, drinking water and sanitation services, all require a dynamic analysis of who does what, at which level, how and with whom, to assess whether governance structures are well-equipped to deliver intended water policy outcomes. This chapter sets the scene and argues that some characteristics exogenous to the water sector, namely size, spatial patterns, demographic dynamics and metropolitan governance arrangements, can affect water management in cities.

Key messages

Assessing the state of play of urban water governance requires understanding the factors affecting water management in cities, be they water-related or not, mapping who does what in water policy design and implementation, and appraising multi-level governance gaps before considering the range of relevant policy responses. This chapter suggests an analytical framework combining these steps while looking at key water functions and diverse situations within and across cities.

Cities in OECD and emerging countries have different features and capacities to respond to water-related challenges. This chapter provides a useful classification of cities facing similar types of challenges in terms of population size, spatial patterns (sprawl, compact), demographic dynamics, or metropolitan governance arrangements.

The current definition of cities reflects their complexity and dynamics while pointing to a more "functional" approach, which for water management is needed to map centres of water demand (urban areas where most people live) and water supply (the surrounding environments where point sources are located) and to consider relevant scales that combine administrative and hydrological considerations.

Why urban water governance matters

Cities are major contributors to national and global sustainable growth building on economic development, well-being and environmental health. The recently-adopted Sustainable Development Goal (SDG) of "making cities inclusive, safe, resilient and sustainable" includes sound water management. As cities will be increasingly exposed to the risks of "too much", "too little" and "too polluted" water over the coming years, the question of which governance frameworks can foster greater resilience and help adapt to changing circumstances is particularly important for cities to prepare for the future and to maintain their central role in local, national and global contexts.

Cities are dynamic environments where several interactions occur. The OECD refers to "cities" as functional urban areas (FUAs), defined by patterns of settlement and human activities and often encompassing multiple municipalities and their commuting zone, whose labour market is highly integrated with the urban cores (Box 1.1; OECD, 2012a). This definition better captures the complexity of cities today and in particular their interconnectedness with their hinterland and other administrative units. In the case of water policy, the functional approach relates to the catchment area where cities are located, which can help map centres of water demand (urban areas where most people live) and water supply (the surrounding environments where point sources are located). Overcoming the mismatch between administrative perimeters and hydrographic boundaries is a question of governance.

Box 1.1. **Redefining "urban" at the OECD**

- **Compact cities**: Dense and proximate development patterns urban areas linked by public transport systems and accessibility to local services and jobs.
- **Functional urban areas** (FUAs): Urban area composed of densely inhabited urban core(s) and hinterland.
- **Metropolitan area**: A functional area with a population between 500 000 and 1.5 million people. A large metropolitan area has a population of 1.5 million or higher.
- **Rural community**: Area characterised by a population density below 150 inhabitants per square kilometre (500 inhabitants for Japan and Korea to account for the fact that the national population density exceeds 300 inhabitants per square kilometre).
- **Urban core**: Highly densely populated contiguous municipalities.
- **Urban hinterland**: Municipalities connected to the urban core by having a certain share of their employed residents working in the urban core.
- **Urban sprawl**: Uncontrolled expansion of urban development characterised by low density, segregated land use and insufficient infrastructure.

Source: OECD (2012a), *Redefining "Urban": A New Way to Measure Metropolitan Areas*, OECD Publishing, Paris, http://dx.doi.org/10.1787/9789264174108-en; OECD (2011a), *OECD Regions at a Glance 2011*, OECD Publishing, Paris, http://dx.doi.org/10.1787/reg_glance-2011-en; OECD (2012b), *Compact City Policies: A Comparative Assessment*, OECD Green Growth Studies, OECD Publishing, Paris, http://dx.doi.org/10.1787/9789264167865-en.

Demographic trends are reshaping local governance, with consequences for urban water management and related investment decisions, amongst others. Such trends can be an opportunity for cities to better articulate urban water policies with broader strategic pathways at local and national levels. Indeed, the global population will reach 9.7 billion people in 2050, 66% of which will be living in cities (UN DESA, 2015). More people will require more water for food, electricity, manufacturing and domestic consumption with an estimated demand rising by 55% globally by 2050. Over the same period, about 4 billion people will be living in water-stressed areas and 20% of the population will be vulnerable to floods, especially in coastal cities (OECD, 2012c).

Fiscal constraints and changing water consumption patterns also bring changes in water demand schedules and the composition of public finance. As competition for water among households, farmers, urban dwellers and industry intensifies, non-state actors, directly or indirectly affected by water policy, will play an increasing role in adaptive water governance. These stakeholders may have the ability to influence water policy outcomes positively or negatively (OECD, 2015a).

Forward-looking water management practices, combining technical (e.g. smart water systems, distributed systems, green infrastructure, etc.) and non-technical innovations (e.g. information systems, water-sensitive urban design) are needed to fit for the future (OECD, 2015b). Technical innovations in urban water management have proven crucial for conserving and reusing water and protecting cities from water-related risks. But resilient urban water management requires both hard and soft infrastructure. This means that institutions play a determinant role in raising awareness; triggering behavioural and policy change; and managing trade-offs across actual users and future generations. This is largely because water crises are primarily *governance* crises that go beyond issues of hydrology, financing and infrastructure.

A city that is resilient from a water management perspective is one that can manage water in a sustainable, integrated and inclusive way, at an acceptable cost, and in a reasonable timeframe as suggested in the OECD Principles on Water Governance (OECD, 2015c). OECD (2015b) argues that cities in OECD countries have not yet solved the problem of water management, and that there is still room for more adaptable infrastructure, different combinations of financing tools and greater protection against present and future water risks, at least cost to society. Such improvements will only happen if water management is adequately governed across multiple scales, authorities, and policy domains. Therefore, policy responses need to be tailored to a given city's needs, while aligning with national goals and priorities.

Methodology

Drawing from the findings of the survey, "Water Governance for Future Cities", carried out across 48 cities in OECD countries and a few emerging economies (Box 1.2), this report explores governance mechanisms that can deal with fragmentation to improve urban water management. This implies looking at co-ordination failures and best practices across water management functions (drinking water supply, sewage collection, wastewater treatment, drainage, water security) and beyond the water chain (e.g. territorial development, spatial planning, biodiversity, ecosystems, waste, energy). The objective is to support policy coherence and effective water management beyond administrative boundaries and sectoral silos. The analytical framework developed in this report seeks to provide insights on *key factors* influencing water governance in cities, *institutional frameworks* to manage water across scales and authorities, *multi-level governance gaps* and *policy responses*, through an evidence-based approach and good practices that can inspire cities surveyed and beyond.

Box 1.2. **OECD Survey on Water Governance for Future Cities**

Cities surveyed

A total of 48 cities form 17 OECD countries and 3 non-OECD countries responded to the survey: Acapulco de Juarez, Amsterdam, Athens, Barcelona, Belo Horizonte, Bologna, Budapest, Calgary, Chihuahua, Cologne, Copenhagen, Culiacan, Daegu, Edinburgh, Glasgow, Grenoble, Hermosillo, Hong Kong, China; Kitakyushu, Krakow, Lisbon, Liverpool, Malaga, Marseille, Mexico City, Milano, Montreal, Nantes, Naples, New York City, Okayama, Oslo, Paris, Phoenix, Prague, Queretaro, Rio de Janeiro, Rome, San Luis Potosi, Singapore, Stockholm, Suzhou, Toluca, Turin, Tuxtla, Veracruz, Zaragoza and Zibo.

The sample covers 15% of the OECD Metropolitan Areas Database, which includes 281 metropolitan areas. An invitation to take part in the survey was extended to 179 cities chosen from the OECD Metropolitan Area Database with a population of more than 500 000 inhabitants. Responses were provided on a voluntary base. The greatest number of responses was received from Mexican (ten), Italian (four) and French cities (four). An overall sample of 32% of EU capital cities is represented. The table below indicates that the surveyed sample, analysed according to key characteristics, shows a good level of representation relative to the OECD Metropolitan Database.

Box 1.2. **OECD Survey on Water Governance for Future Cities** *(continued)*

Representativeness of the sample surveyed relative to the OECD Metropolitan Areas Database

Cities	Sample surveyed (42*)	OECD Metropolitan Areas database (281)	Over/under representation
More than 5 million	7%	6%	1.3
Between 1.5 million and 5 million	36%	19%	1.9
Less than 1.5 million	64%	75%	0.9
Higher density than the OECD metropolitan areas population density median	55%	50%	1.1
Lower density than the OECD metropolitan areas population density median	45%	50%	0.9
Above the OECD metropolitan areas population average growth rate	44%	45%	1.0
Below the OECD metropolitan areas population average growth rate	56%	55%	1.0

*Cities from non-OECD countries (six in total) are not included.

Methodological note

The OECD Survey was pilot-tested by a few cities before it was launched on line on 8 August 2014 in English, French, Italian and Spanish. It was also translated into Chinese and Japanese to facilitate responses from these two countries. The selection of cities from the OECD Metropolitan Areas Database sought to facilitate comparisons in terms of the relationship between water-related and socio-economic variables in urban areas (total population, population density, gross domestic product [GDP], etc.) building on data available at the OECD. A few cities from non-OECD countries were also included in the sample in order to investigate water-related challenges and policy responses in diverse, yet comparable, contexts.

The respondents targeted were the primary authorities managing water in cities. However, the nature of respondents varies in line with the core responsibilities for water management across and within countries covered in the sample: e.g. municipal or metropolitan departments, regional authorities, service providers or deconcentrated (administrative) bodies at the local level. Fact-based and perception-based data were provided by highly qualified technical and administrative experts operating in any of above authorities of surveyed cities. In principle, the questionnaire required co-ordination between different actors dealing with water resources and water services in surveyed cities.

The survey requested data for 1990, 2000, and 2012. In many cases, data for 1990 were not available. In particular, a full analysis of revenues and expenditure for the water sector at city level was not possible due to incomplete data on sources. Data were collected from respondents on the basis of their knowledge and available sources of information. It has to be noted that some indicators (e.g. water loss, population in households connected to wastewater treatment) are regularly collected in some OECD countries, but a common methodology on these statistics has proven difficult to establish thus far. This caveat should be taken into account when considering cities' performance on the basis of the above-mentioned indicators.

Structure of the questionnaire

- Section 1: Factors having an impact on water governance in the city and its surrounding
- Section 2: Key indicators and trends related to water resources and services
- Section 3: Urban water financing (revenues and expenditures)
- Section 4: Mapping who does what
- Section 5: Stakeholder engagement in water-related decision-making
- Section 6: Linkages between the city and its surroundings (e .g. rural areas, urban hinterlands)
- Section 7: Evaluation tools to assess urban water management
- Section 8: Multi-level governance gaps in urban water management
- Section 9: Forward-looking and adaptive strategies to cope with future water challenges

Analytical framework

The report proposes an analytical framework (Figure 1.1) that combines: i) an assessment of key factors affecting the effectiveness of urban water governance; ii) a mapping of who does what at which level; iii) an appraisal of the main multi-level governance gaps to urban water management; and iv) a zoom on policy responses to mitigate fragmentation and foster integrated urban water management in cities and their hinterlands.

Figure 1.1. **The analytical framework for assessing water governance in cities**

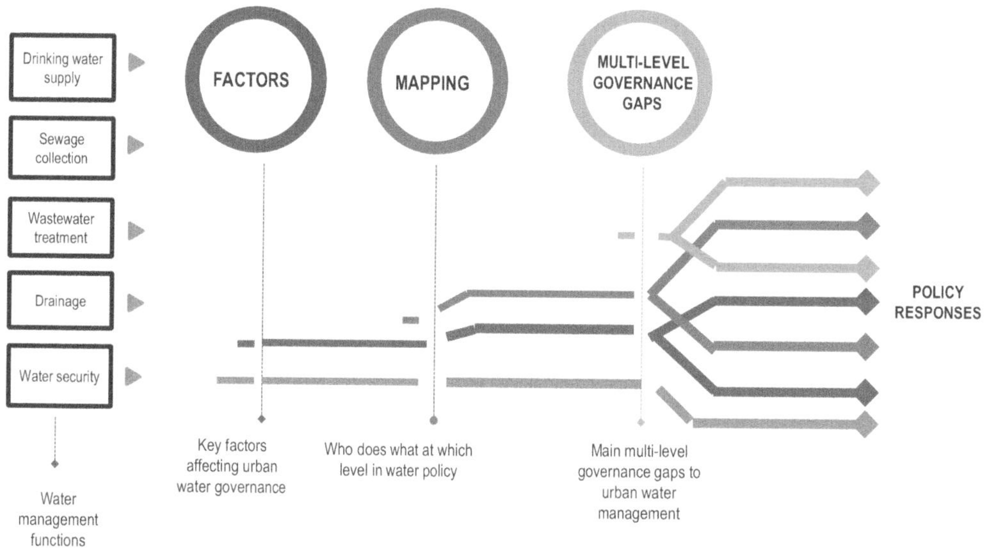

Source: Author's elaboration.

The policy guidance provided in this report is articulated around the "3Ps" co-ordination framework across *policies, people and places* (Figure 1.2) for several reasons. First, water governance has consequences on, and can be affected by, a number of intrinsically related *policies* such as land use, spatial planning, transport, energy, solid waste, environment, and agriculture. Similarly, a plethora of *people* from public, private, non-profit sectors to water users themselves have a stake or play a role in urban water management. Finally, water boundaries cut across *places* in terms of cities (i.e. when concerning more municipalities in a metropolitan area) or hinterland (i.e. the surrounding environment, rural areas and watersheds, which sustain the major bulk of water demand from cities and where the actual sources of water are often located). Particular emphasis is put on the most prominent governance mechanisms that can foster effectiveness, efficiency and inclusiveness of urban water governance building on the OECD Principles on Water Governance (OECD, 2015c). They relate to vertical and horizontal co-ordination across *policies*, dedicated metropolitan arrangements and rural-urban partnerships (co-ordination across *places*), and stakeholder engagement (co-ordination across *people*). Their selection and implementation depends on local specificities as well as on an evaluation of their costs and benefits.

Figure 1.2. **The 3Ps co-ordination framework for integrated urban water management**

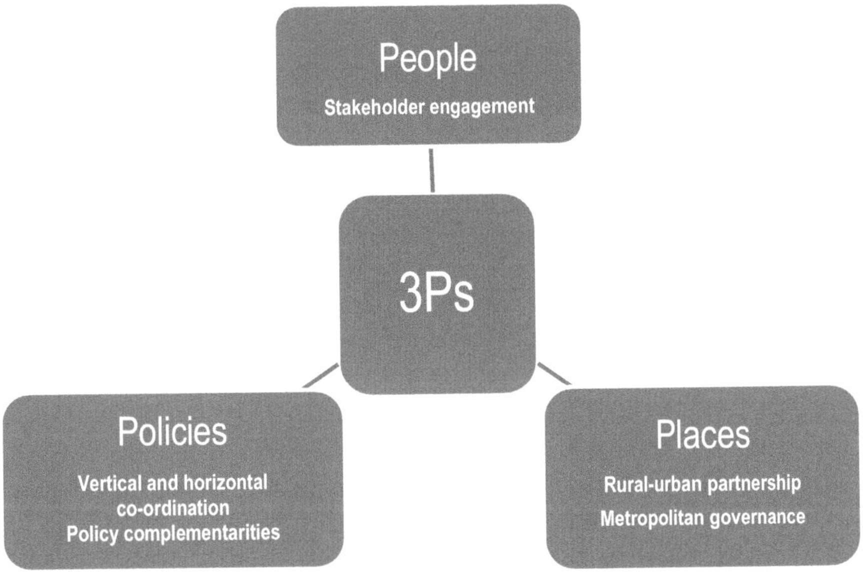

Source: Author's elaboration.

The report argues that cities on their own cannot cope with water challenges and that co-ordination across local authorities and central governments is absolutely crucial. A number of policy conclusions can be drawn to foster stronger urban-national policy frameworks that can help to better manage too much, too little and too polluted water in cities and their hinterland. Three policy questions deserve particular attention:

1. What can be done at the local level to prepare urban water governance to the future?
2. What should cities do in co-operation with their hinterland?
3. How to enhance national/local co-operation on urban water governance?

As urban water governance is a shared responsibility across policy makers and multiple stakeholders, the question of co-ordination is essential. However, equal attention needs to be paid to the trade-offs that such co-ordination efforts imply as it takes time and (institutional) effort, and it can generate multiple types of costs. Given that cities often seek value for money, the cost dimension needs to be factored in and compared with benefit. Moreover, being governance a means to an end, the level and type of governance structures and mechanisms has to match the level of risks. That means for instance that light co-ordination mechanisms may be more cost-beneficial than more complex ones.

Characteristics of cities affecting urban water governance

Several studies define cities according to their objectives and/or characteristics (e.g. green cities, smart cities, compact cities etc.), for example:

- Brown, Keath and Wong (2008), identify a framework for the development of urban water transitions policy based on a study across Australian cities, which recognises the temporal changes of cities towards a "water sensitive" model. This kind of "water-sensitive city" approach requires environmental sustainability, but also a reform of the existing contract between citizens and governments over water policy, infrastructure, technologies and urban form (OECD, 2011b).
- The European Union (EU) project on **Tr**ansitions to the **U**rban Water **S**ervices of **T**omorrow (TRUST)[1] identifies green cities, water scarcity regions and urban/peri-urban metropolitan areas to test the implementation of innovative solutions for a more sustainable water future.
- The European Innovation Partnership EIP Water through the City Blueprint Action Group[2] assesses the sustainability of water management in cities and regions, linking with the European Innovation Partnership on Smart Cities. The project aims to encourage cities to become "smarter" by: i) designing a long-term social, economic and ecological agenda; ii) implementing a circular economy focussing on social innovation and better governance; iii) exploring co-benefits for cities in co-ordinating across sectors.

Clustering cities that face similar types of characteristics can favour a city-to-city learning process, fostering an exchange of best practices across cities facing common patterns or seeking to achieve similar goals. This report proposes to cluster cities by size, spatial patterns, demographic dynamics and metropolitan governance arrangements (Figure 1.3) along the following lines:

- more than 5 million inhabitants, between 1.5 million and 5 million inhabitants and less than 1.5 million inhabitants (cities by size)
- lower or higher population density compared to the OECD metropolitan areas population density median (cities by spatial patterns)
- population growth rate above or below the OECD metropolitan areas average (cities by demographic dynamics)
- existence of (formal or informal) metropolitan governance arrangements or not (cities by metropolitan governance arrangements).

Clustering cities according to specific characteristics helps to account for the different challenges cities face and propose place-based responses. In the first place, the geographic position of cities determines the main challenges they are exposed to as well as their capacity to respond due to possible physical constraints (e.g. delta cities will differ from those located in mountainous areas, in terms of water-related risks they need to tackle). In addition, features such as the size, spatial organisation, demographic dynamics and governance structures of a city have an impact on water functions. Whether a city is big or small, compact or sprawl, growing or shrinking affects water consumption patterns, utilities' management models, linkages across sectors, ability to engage stakeholders and overall implementation capacity.

Figure 1.3. **Types of cities according to key characteristics**

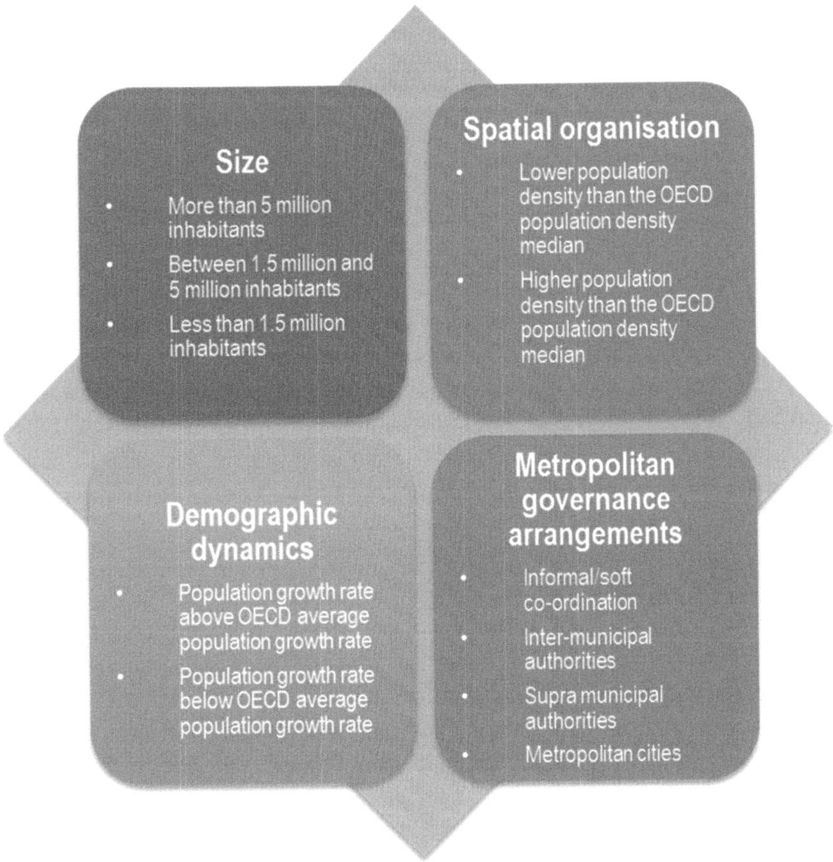

Source: Author's elaboration.

Cities by size

Competition across water uses stemming from the diverse needs of large conurbations and megacities has an impact on water quality and quantity. A total of 12% of the world's urban population is concentrated in the 28 current megacities (UN DESA, 2014) mainly located in Asia, but also in Latin America, Africa, Europe and North America. The increasing number of megacities (13 more by 2030, most of which will also be located in Asia; UN DESA, 2014) will raise the need for expanded water and sanitation services. This has implications for the type of infrastructure needed and the kind of utilities management model applied. Water quality is also impaired by urban and industrial activities generating water pollution. Moreover, large conurbations and megacities surveyed in this report (e.g. Hong Kong, China; Mexico City, New York, Rio de Janeiro, Singapore; Table 1.1) are either cities at risk of flood or subject to increasing droughts. These risks are compounded by climate change and call for greater attention to water security.

Table 1.1. **Surveyed cities by size**

No. of inhabitants	Cities
More than 5 million	Mexico City, New York City, Paris, Hong Kong,China; Rio de Janeiro, Singapore
Between 1.5 million and 5 million	Amsterdam, Athens, Barcelona, Belo Horizonte, Budapest, Daegu, Lisbon, Marseille, Milan, Montreal, Naples, Phoenix, Rome, Suzhou, Zibo
Less than 1.5 million	Acapulco, Bologna, Calgary, Chihuahua, Cologne, Copenhagen, Culiacan, Edinburgh, Glasgow, Grenoble, Hermosillo, Kitakyushu, Krakow, Liverpool, Malaga, Nantes, Okayama, Oslo, Prague, Queretaro, San Luis Potosi, Stockholm, Turin, Toluca, Tuxtla, Veracruz, Zaragoza

Source: Based on population of the city area - FUA (OECD, 2012d) and data provided by surveyed cities from non-OECD countries.

Medium and smaller-sized cities are also important for economic growth and sustainable urban development. According to UN DESA (2014) almost 50% of the world's urban population lives in settlements of less than 500 000 inhabitants. Over the next 15 years, small and medium cities are expected to generate nearly 40% of the global economic growth in emerging economies, including those located in Asia (Dobbs et al., 2011). This category of cities must build sufficient governance capacity to ensure policy integration across water and related sectors and to secure adequate financing to implement necessary measures.

Spatial organisation

The spatial organisation of a city has an impact on water consumption trends, infrastructure development and water footprint. During the last decade, cities in OECD countries have experienced suburbanisation trends with population growth in commuting zones growing faster than in core areas (Veneri, 2015); emerging countries have been facing intense phenomena of urban sprawl (Kamal-Chaoui and Robert, 2009). Typically, sprawl cities put greater pressure on the environment than compact cities due to land use stress, fragmentation of natural habitats and increasing air pollution emissions (Cirilli and Veneri, 2014). While they are likely to generate greater maintenance costs, they may be beneficial when it comes to rainwater drainage, groundwater recharge, and flood and scarcity management.

Compact cities may require a relatively low level of infrastructure construction for water and sewer systems and are likely to increase the efficiency of infrastructure investment, while reducing the cost of maintenance of water supply systems. However, they are less likely to accommodate green infrastructure, hence retrofitting existing urban water infrastructures can be more costly (OECD, 2015b). Generally, compact cities tend to provide citizens with easier access to a diversity of local services (OECD, 2012b). More compact development has been encouraged since the 1990s by the European Commission and countries such as the Netherlands and United Kingdom have identified "compactness" as a crucial element for sustainable development policies (OECD, 2012c). Exploring opportunities for synergies between spatial planning and wastewater management to promote a long-term view can be beneficial for both water quantity and quality. See Table 1.2 for the list of cities surveyed for this report by spatial organisation.

Table 1.2. **Surveyed cities by spatial organisation**

Population density	Cities
Lower density	Acapulco, Bologna, Budapest, Calgary, Chihuahua, Culiacan, Grenoble, Hermosillo, Krakow, Marseille, Montreal, Nantes, Oslo, Phoenix, Prague, San Luis Potosi, Stockholm, Tuxtla, Zaragoza
Higher density	Amsterdam, Athens, Barcelona, Belo Horizonte, Cologne, Copenhagen, Daegu, Edinburgh, Glasgow, Hong Kong, China; Kitakyushu, Lisbon, Liverpool, Malaga, Mexico City, Milan, Naples, New York City, Okayama, Paris, Querétaro, Rio de Janeiro, Rome, Singapore, Suzhou, Toluca, Turin, Veracruz, Zibo

Note: Cities are classified according to their position below or above the OECD Metropolitan Areas population density median: 481 persons per square kilometre. Population density is the ratio between total population of the functional urban area and total land area (OECD, 2012d). For Suzhou and Zibo, data relate to the provinces of Jiangsu and Shandong, respectively, Worldatlas website (2015), Province Of Jiangsu and Province Of Shandong, China, www.worldatlas.com, (accessed November 2015). Other sources were used for Belo Horizonte (UN Habitat, 2010), Information Services Department (2015), Hong Kong Special Administrative Region Government, Hong Kong: the facts, www.gov.hk, (accessed December 2015); Department of statistics Singapore (2015), Singapore in Figures 2015, www.singstat.gov.sg, (accessed December 2015); and WPR website, Rio de Janeiro Population 2015, www.worldpopulationreview.com, (accessed December 2015).

Demographic dynamics

Demographic dynamics also affect water demand and supply and can challenge the capacity of local governments to manage water resources efficiently in the face of environmental degradation and economic trends. Shrinkage and expansion trends have an influence on water infrastructure needs through changes in the demand schedule and by altering the composition of public finances as the base shrinks or expands. Adaptation to these trends should be reflected in investment choices and management models for supplying services.

Urban shrinkage characterised by declining population is a consistent trend in many OECD countries.[3] For instance, the Japanese population is expected to contract from 127 million in 2010 to 95 million by 2050 (Population Reference Bureau, 2010). In Germany and Italy the population is ageing and the fertility rate is low (OECD, 2015d). Among the sample of 48 cities surveyed, 23 reported having a population growth rate above the OECD average (0.8), and 25 reported a population growth rate below the OECD average. Of the 25 with a growth rate below the OECD average, 4 cities (Athens, Budapest, Kitakyushu and Krakow) reported negative rates, showing a declining population between 2000 and 2012.

The economic effects of a shrinking population are generally loss in tax bases and revenues on the one hand, and increase in public expenditure to maintain infrastructure on the other hand (Pallagst, Wiechmann and Martinez-Fernandez, 2010). Urban shrinkage may also have implications in terms of decreasing water consumption, leading to increased prices of water and wastewater, as the fixed costs of infrastructure provision have to be borne by fewer people (Naumann and Wissen, 2006). Reduced flow quantity can impair proper functioning of supply networks and have consequences on the cost of the oversized infrastructure, which is usually fixed. As the population shrinks, the size of existing water networks may be too large in relation to adjusted demand. This may call for alternative management models (e.g. smart distributed systems) that can provide greater room for adjustment to changing dynamics. How to favour a transition toward new types of models is a question to which water managers will need to respond.

Demographic expansion and the speed at which some cities are growing put pressure on water resources. For example, cities in Asia are dealing with demographic growth, urban built area and GDP growth, which are likely to intensify competition across water users; several fast-growing cities in the United States are located in dry areas of the south (OECD, 2015b). The majority of surveyed Mexican cities, often affected by water scarcity issues, grew above the OECD average city population growth rate between 2000 and 2012. It is important to note that a number of surveyed cities showing population growth lower than the OECD average still grew during the same timeframe (Table 1.3).

Table 1.3. **Surveyed cities by demographic dynamics**

Population growth	Cities
Below OECD average population growth rate	Athens, Belo Horizonte, Bologna, Budapest, Cologne, Copenhagen, Daegu, Glasgow, Grenoble, Hong Kong, China; Kitakyushu, Krakow, Lisbon, Liverpool, Marseille, Milan, Nantes, Naples, New York City, Okayama, Paris, Prague, Rome, Stockholm, Turin, Zibo
Above OECD average population growth rate	Acapulco, Amsterdam, Barcelona Calgary, Chihuahua, Culiacan, Edinburgh, Hermosillo, Malaga, Mexico City, Montreal, Oslo, Phoenix, Queretaro, Rio de Janeiro, San Luis Potosi, Singapore, Suzhou, Toluca, Tuxtla, Veracruz, Zaragoza

Note: Cities are grouped according to whether their population growth rates are below or above the OECD functional urban area's population average growth rate (0.8) years 2000-12. Based on total population (OECD, 2012d), *OECD Regional Statistics: Metropolitan areas* (Edition 2015), year 2012 and 2000, *OECD Regional Statistics* (database), http://dx.doi.org/10.1787/a06d83f9-en (accessed November 2015), and data provided by surveyed cities from non-OECD countries.

Cities by metropolitan governance arrangements

OECD countries face a consistent trend towards the creation of metropolitan governance bodies in which public authorities' responsibilities are organised at the metropolitan level (OECD, 2015e). In some cases, the metropolitan authority has some responsibility over the water sector. But overall, metropolitan bodies, regardless of their sectoral competences, can help to improve service organisation, pool infrastructure investment and capacity, and co-ordinate policies at the relevant scale. OECD (2015f) emphasised the costs and benefits of agglomerations. In particular, the report suggested that increasing fragmentation of a metropolitan area into different municipalities is correlated to lower levels of labour productivity, which well-placed metropolitan governance structures may mitigate. Beyond dedicated metropolitan governance structures, there is a wide range of metropolitan governance arrangements (informal and formal) where co-operation between cities can foster synergies and complementarities in terms of roles and responsibilities (OECD, 2015f).

See Table 1.4 for the list of cities surveyed for this report by metropolitan governance arrangements and Table 1.5 for a sample of cities surveyed grouped according to similar features.

Table 1.4. **Surveyed cities by metropolitan governance arrangements**

Metropolitan governance arrangements	Cities
Metropolitan governance arrangements in place	Acapulco, Amsterdam, Barcelona, Belo Horizonte, Bologna, Budapest, Calgary, Chihuahua, Cologne, Copenhagen, Culiacan, Daegu, Edinburgh, Glasgow, Grenoble, Hermosillo, Kitakyushu, Lisbon, Marseille, Mexico City, Milan, Montreal, Nantes, Naples, Paris, Phoenix, Queretaro, Rome, San Luis Potosí, Stockholm, Suzhou, Toluca, Turin, Tuxtla, Veracruz
Metropolitan governance arrangements not in place	Athens, Hong Kong, China; Krakow, Liverpool, Malaga, New York City, Okayama, Oslo, Prague, Rio De Janeiro, Singapore, Zaragoza, Zibo

Note: These governance arrangements are not necessarily specific to the water sector, but reflect the approach to multi-level co-ordination of the city and its surroundings.

Source: Based on the classification provided in OECD (2015f), which distinguished among: informal/soft co-ordination (lightly institutionalised platforms for information sharing and consultation); inter-municipal authorities (single or multi-purpose authorities); supra-municipal authorities (an additional layer above municipalities); metropolitan cities (special status putting on the same footing as the next upper level of government) and the results of the OECD Metropolitan Governance Survey (Ahrend, R., C. Gamper and A. Schumann, 2014). In addition, desk research was carried out for the cities of Budapest, Stockholm, Milan, Rome, Naples and Turin, which were not included in the above-mentioned study. Cities where "metropolitan governance arrangements are not in place" are those for which none of the metropolitan governance arrangements were listed in the OECD Metropolitan Governance Survey or that responded "No" to the question: "Does your city belong to a metropolitan governance body?" of the OECD Survey on Water Governance for Future Cities (2014).

Table 1.5. **Sample of surveyed cites according to similar features**

Cities	Size (Population of the city area-FUA)	Demographic dynamics (Population growth rate below or above the OECD average)	Spatial patterns (Population density higher/lower than the OECD median)	Existence (and type) of metropolitan governance arrangements
Singapore Rio de Janeiro				No metropolitan arrangement
Hong Kong, China New York City				No metropolitan arrangement
Mexico City				
Paris				
Barcelona Amsterdam				
Milan Rome Naples				
Lisbon Belo Horizonte				
Marseille Budapest				
Montreal				
Athens Zibo				No metropolitan arrangement

Table 1.5. **Sample of surveyed cites according to similar features** *(continued)*

Cities	Size (Population of the city area-FUA)	Demographic dynamics (Population growth rate below or above the OECD average)	Spatial patterns (Population density higher/lower the OECD median)	Existence (and type) of metropolitan governance arrangements
Phoenix				
Daegu		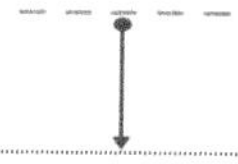		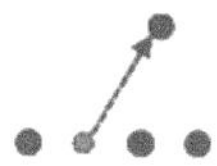
Suzhou			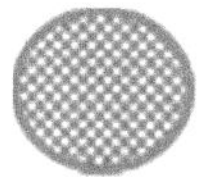	Existence of metropolitan arrangements, type unknown
Calgary				
Copenhagen Turin		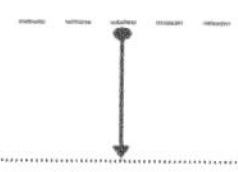	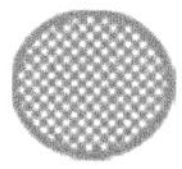	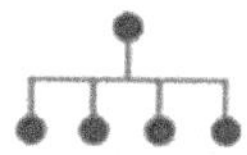
Acapulco Chihuahua Culiacan San Luis Potosi Tuxtla Hermosillo		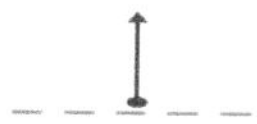		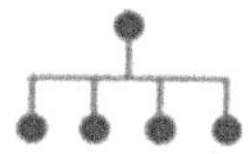
Queretaro Veracruz Toluca				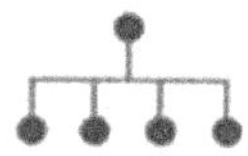
Nantes Grenoble Stockholm		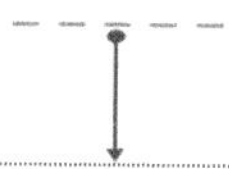		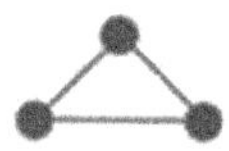
Oslo Zaragoza				No metropolitan arrangement
Malaga			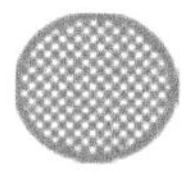	No metropolitan arrangement

Table 1.5. **Sample of surveyed cites according to similar features** *(continued)*

Cities	Size (Population of the city area-FUA)	Demographic dynamics (Population growth rate below or above the OECD average)	Spatial patterns (Population density higher/lower the OECD median)	Existence (and type) of metropolitan governance arrangements
Edinburgh				
Glasgow Kitakyushu Cologne	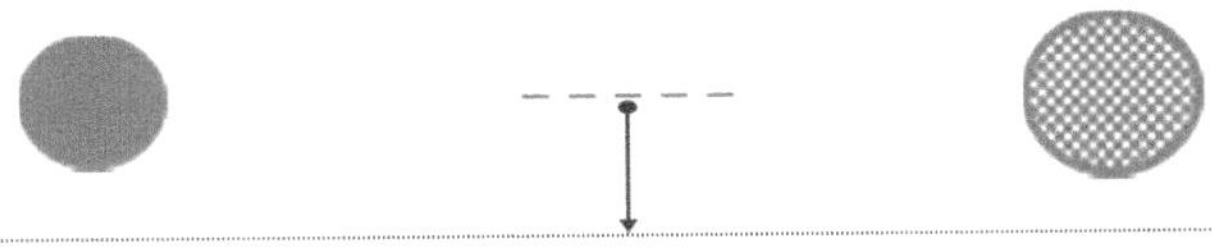			
Liverpool Okayama	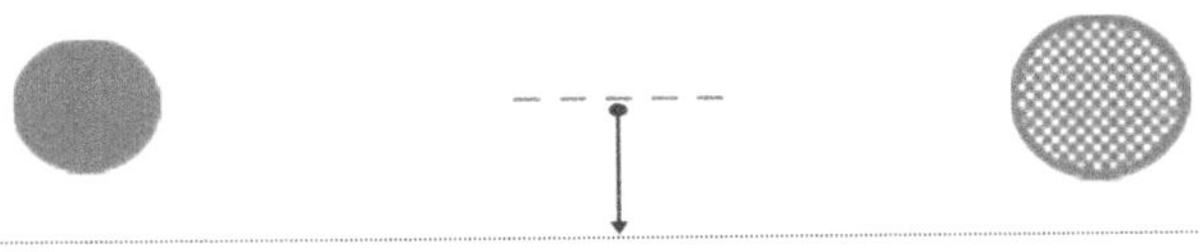			No metropolitan arrangement
Bologna	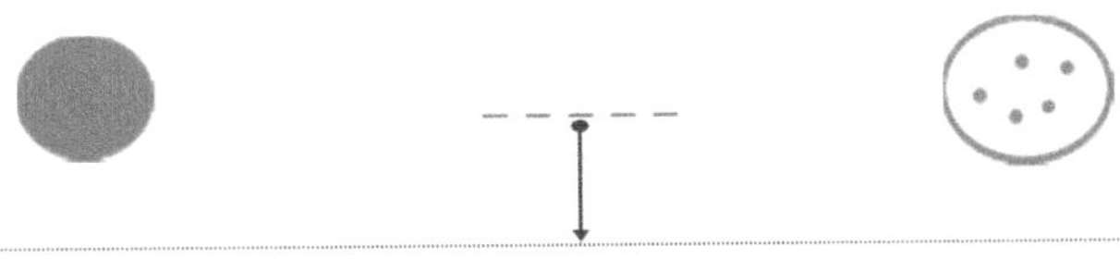			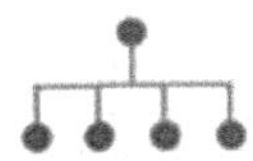
Krakow Prague	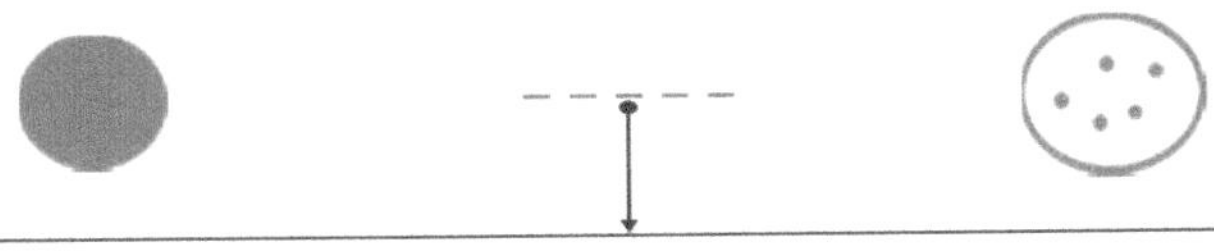			No metropolitan arrangement

Legend

Size: Population of the urban core of the functional urban area	
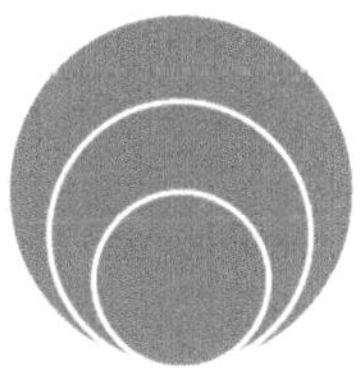	More than 5 million inhabitants Between 1.5 million and 5 million inhabitants Less than 1.5 million inhabitants
Demographic dynamics: Population growth rate of the functional urban area	
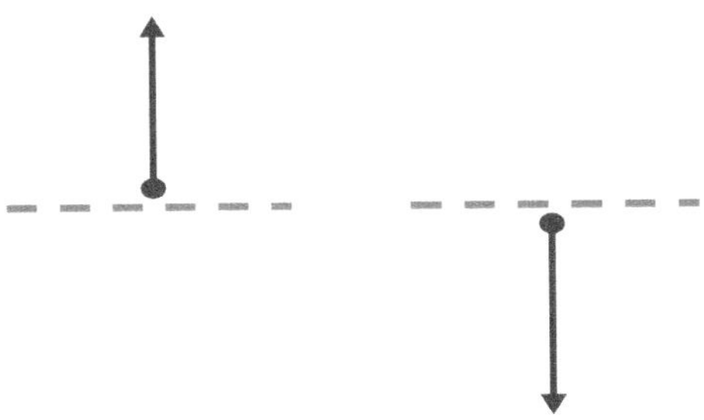	Above OECD population average growth rate Below OECD population average growth rate

Spatial patterns: Population density of the functional urban area	
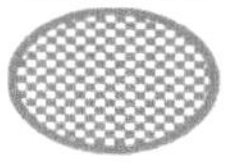	Higher than OECD population density median
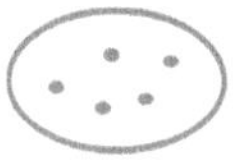	Lower than OECD population density median
Metropolitan governance arrangements	
	Informal/soft co-ordination
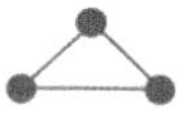	Inter-municipal authority
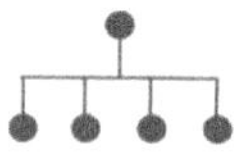	Supra-municipal authority
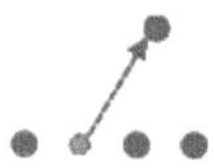	Metropolitan city

Source: Author's elaboration.

Notes

1. See TRansitions to the Urban Water Services of Tomorrow (TRUST) website (n.d.), www.trust-i.net/, (accessed November 2015).

2. See EIP Water website (n.d.), CITY BLUEPRINTS - Improving Implementation Capacities of Cities and Regions, http://www.eip-water.eu/City_Blueprints#sthash.XHkjP6Tb.dpuf, (accessed December 2015).

3. Differences in demographic dynamics can be observed in urban and rural contexts. In OECD countries, rural population shrank throughout the second half of the 20th century, while urban population grew (OECD/China Development Research Foundation, 2010). However, empirical evidence shows that the proximity to urban centres can have positive impacts on the population growth rate of rural regions, as they can potentially benefit from better access to services and markets (Veneri and Ruiz, 2013; OECD, 2015c).

Bibliography

Ahrend, R., C. Gamper and A. Schumann (2014), "The OECD Metropolitan Governance Survey: A quantitative description of governance structures in large urban agglomerations", *OECD Regional Development Working Papers*, No. 2014/04, OECD Publishing, Paris, http://dx.doi.org/10.1787/5jz43zldh08p-en.

Brown R., N. Keath and T. Wong (2008), "Transitioning to water sensitive cities: Historical, current and future transition states", 11th International Conference on Urban Drainage, Edinburgh, Scotland, UK, available at: http://web.sbe.hw.ac.uk/staffprofiles/bdgsa/11th_International_Conference_on_Urban_Drainage_CD/ICUD08/pdfs/618.pdf.

Cirilli, A. and P. Veneri (2014), "Spatial structure and carbon dioxide (CO2) emissions due to commuting: An analysis of Italian urban areas", *Regional Studies*, Volume 48/12.

Department of statistics Singapore (2015), Singapore in Figures 2015, www.singstat.gov.sg, (accessed December 2015).

Dobbs, R., et al. (2011), *Urban World: Mapping the Economic Power of Cities*, McKinsey Global Institute.

EIP Water website (n.d.), CITY BLUEPRINTS - Improving Implementation Capacities of Cities and Regions, http://www.eip-water.eu/City_Blueprints#sthash.XHkjP6Tb.dpuf, (accessed December 2015).

Information Services Department (2015), Hong Kong Special Administrative Region Government, Hong Kong: the facts, www.gov.hk, (accessed December 2015).

Kamal-Chaoui, L. and A. Robert (2009), "Competitive cities and climate change", *OECD Regional Development Working Papers*, No. 2009/02, OECD Publishing, Paris, http://dx.doi.org/10.1787/218830433146.

Naumann, M. and M. Wissen (2006), "Infrastructural commercialization and uneven development: The case of East Germany and Matthias", in IHP/UNESCO, *Urban Water Conflicts: An Analysis of the Origins and Nature of Water-related Unrest and Conflicts in the Urban Context*, UNESCO Working Series SC-2006/WS/19.

OECD (2015a), *Stakeholder Engagement for Inclusive Water Governance*, OECD Studies on Water, OECD Publishing, Paris, http://dx.doi.org/10.1787/9789264231122-en.

OECD (2015b), *Water and Cities. Ensuring Sustainable Futures*, OECD Studies on Water, OECD Publishing, Paris, http://dx.doi.org/10.1787/9789264230149-en.

OECD (2015c), "OECD Principles on Water Governance", brochure, OECD, Paris, available at: www.oecd.org/gov/regional-policy/OECD-Principles-on-Water-Governance-brochure.pdf.

OECD (2015d), *Ageing in Cities*, OECD Publishing, Paris, http://dx.doi.org/10.1787/9789264231160-en.

OECD (2015e), *The Metropolitan Century: Understanding Urbanisation and its Consequences*, OECD Publishing, Paris, http://dx.doi.org/10.1787/9789264228733-en.

OECD (2015f), *Governing the City*, OECD Publishing, Paris, http://dx.doi.org/10.1787/9789264226500-en.

OECD (2012a), *Redefining "Urban": A New Way To Measure Metropolitan Areas*, OECD Publishing, Paris, http://dx.doi.org/10.1787/9789264174108-en.

OECD (2012b), *Compact City Policies: A Comparative Assessment*, OECD Green Growth Studies, OECD Publishing, Paris, http://dx.doi.org/10.1787/9789264167865-en.

OECD (2012c), *OECD Environmental Outlook to 2050: The Consequences of Inaction*, OECD Publishing, Paris, http://dx.doi.org/10.1787/9789264122246-en.

OECD (2012d), "Metropolitan areas", *OECD Regional Statistics* (database), http://dx.doi.org/10.1787/data-00531-en.

OECD (2011a), *OECD Regions at a Glance 2011*, OECD Publishing, Paris, http://dx.doi.org/10.1787/reg_glance-2011-en.

OECD (2011b), *Water Governance in OECD Countries: A Multi-level Approach*, OECD Studies on Water, OECD Publishing, Paris, http://dx.doi.org/10.1787/9789264119284-en.

OECD/China Development Research Foundation (2010), *Trends in Urbanisation and Urban Policies in OECD Countries: What Lessons for China?*, OECD Publishing, Paris, http://dx.doi.org/10.1787/9789264092259-en.

Pallagst, K., T. Wiechmann and C. Martinez-Fernandez (eds.) (2010), *Shrinking Cities: International Perspectives and Policy Implications*, Routledge, New York.

Population Reference Bureau (2010), "2010 World Population Data Sheet", available at: www.prb.org/pdf10/10wpds_eng.pdf.

TRansitions to the Urban Water Services of Tomorrow (TRUST) website (n.d.), www.trust-i.net/, (accessed November 2015).

UN DESA (2015), "Revision of World Population Prospects", www.un.org/development/desa/en/news/population/2015-report.html, (accessed December 2015).

UN DESA (2014), "Revision of World Population Prospects", https://esa.un.org/undp/wup, (accessed December 2015).

UNHABITAT (2010), *Solid Waste Management in the World's Cities. Water and Sanitation in the World's Cities*, Earthscan, London and Washington, United Nations Human Settlements Programme.

Veneri, P. (2015), "Urban spatial structure in OECD cities: Is urban population decentralising or clustering?", *OECD Regional Development Working Papers*, 2015/01, OECD Publishing, Paris, http://dx.doi.org/10.1787/5js3d834r3q7-en.

Veneri, P. and V. Ruiz (2013), "Urban-to-rural population growth linkages: Evidence from OECD TL3 regions", *OECD Regional Development Working Papers*, 2013/03, OECD Publishing, Paris, http://dx.doi.org/10.1787/ 5k49lcrq88g7-en.

Worldatlas website (2015), Province Of Jiangsu and Province Of Shandong, China, www.worldatlas.com, (accessed November 2015).

WPR website, Rio de Janeiro Population (2015), http://www.worldpopulationreview.com, (accessed December 2015).

Chapter 2

Factors shaping urban water governance

This chapter analyses key factors affecting water management in cities. These endogenous and exogenous factors shape urban governance by requiring adaptation to changing circumstances, in terms of capacities, data collection, information disclosure and stakeholder engagement, amongst others. The chapter argues that understanding the factors shaping urban water governance can help to devise more effective answers and determine priorities, and concludes with an overview of the challenges cities are facing or are likely to face according to their features by size, spatial patterns, demographic dynamics and metropolitan governance.

Key messages

Several factors have an impact on resilience of cities in terms of managing water risks now and in the future. They can be endogenous and exogenous to the water sector, but hold policy implications both now and in the future in terms of who does what, at which level, and how.

While cities in OECD countries currently provide high quality water services, infrastructure renewal, upgrade and maintenance pose serious concerns and require managing a number of trade-offs in terms of who pays for what across current and future generations, rural and urban areas and categories of users.

Greater awareness is needed as water risks are likely to have a strong impact on citizens' well-being and sustainable growth. Climate change and urban growth have consequences on water governance. While some cities have set up adaptive strategies and made significant progress (e.g. reduction in per capita domestic water consumption, decrease in water leakage and greater access to services), others are lagging behind and need to catch up.

Different cities face different challenges and have different capacities to respond. Cities' features in terms of size, spatial patterns, demographic dynamics and metropolitan governance have an impact on water consumption patterns and management models, as they influence linkages across sectors as well as the ability to engage stakeholders and properly implement policies.

Institutional, legal and regulatory factors (e.g. territorial and administrative reforms) also have an influence on water management in cities by creating new forms of interactions between institutions, places and sectors.

Cities may not face all problems simultaneously; therefore, a temporal dimension is needed to distinguish between short-, medium- and long-term factors and to qualify the urgency of responses accordingly.

Water infrastructure challenges in cities

Almost all surveyed cities (92%) signalled infrastructure as a major factor in urban water governance (Figure 2.1). It is estimated that investment requirements in the water supply and treatment sector of OECD countries will increase by almost 50% by 2030 (OECD, 2007). Over the period 2013-30, countries representing 90% of global gross domestic product (GDP) will face water infrastructure costs of USD 11.7 trillion, which amounts to more than for telecommunication and less than for roads (McKinsey, 2013 in WWC and OECD, 2015). When asked to associate the top 5 words, from a choice of 65, with water management in cities, respondents gave a high relevance to "infrastructure" and "efficiency" (Figure 2.2). Empirical research shows that heavy investment in water efficiency could boost overall GDP and create jobs (Hammer et al., 2011). A range of innovative options can increase efficiency and save costs, from sophisticated information and communication technology (ICT) tools for monitoring water service operations to energy costs minimisation for wastewater treatment.

Figure 2.1. **Factors affecting urban water governance**

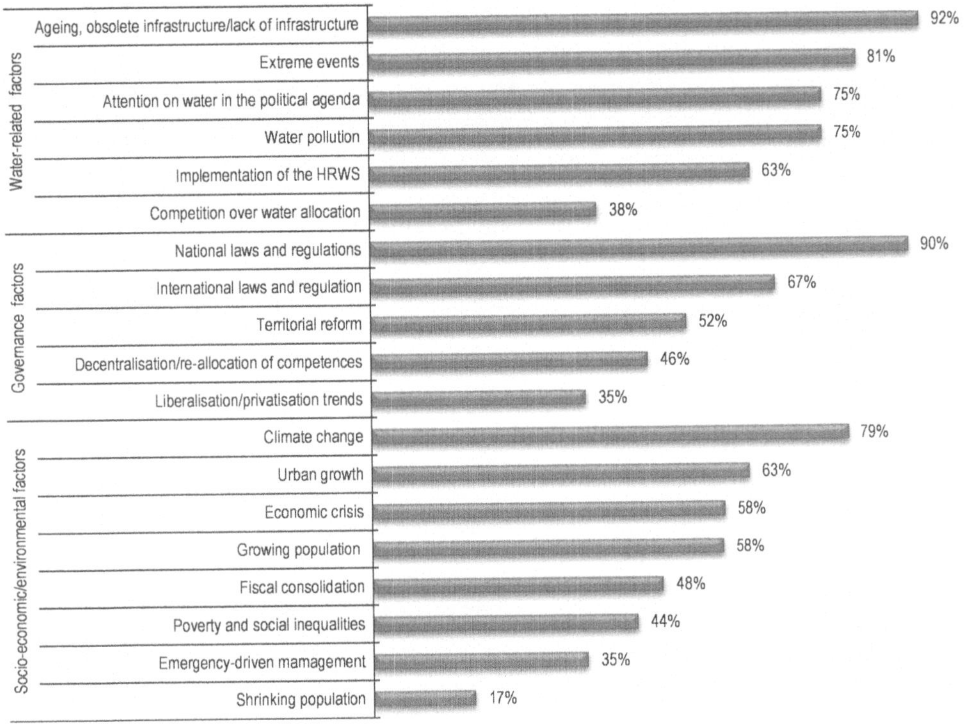

Note: "Critical" or "important" factors based on the responses provided by 48 surveyed cities on a four-scale rating from "critical" to "not important".

Source: OECD Survey on Water Governance for Future Cities (2014).

Figure 2.2. **Keywords associated with "water management in cities"**

Note: Words selected out of 65 options and ranked first on a scale from 1 to 5.

Source: OECD Survey on Water Governance for Future Cities (2014).

Ageing water networks have negative impacts in terms of efficiency. One of the consequences is water loss. On average, water loss in surveyed cities was 21% in 2012. In the Mexican cities surveyed, however, figures reported were higher than 40% (Chihuahua, Mexico City, San Luis Potosi) or even 60% (Tuxtla) (Figure 2.3). Smaller cities surveyed (under 1.5 million inhabitants) reported higher average water loss than larger cities. The correlation between GDP per capita and the share of water losses shows greater water losses in cities with lower GDP per capita (Figure 2.4). However the low value of the R^2 shows that, in this case, water losses are not well explained by looking only at differences in GDP per capita. Wastage in these cities is generally associated with unauthorised consumption, poor connections and metering inaccuracies (Farley, 2001). The economically optimal level of water losses in municipal networks is estimated to be on average between 10% and 20%, depending on the nature of individual systems (OECD, 2006a). According to an approach adopted by the European Commission , this level is reached at the point at which the cost of reducing leakage is equal to the benefit gained from further leakage reductions (European Commission, 2013).

Figure 2.3. **Proportion of water loss in surveyed cities (leakage rate)**

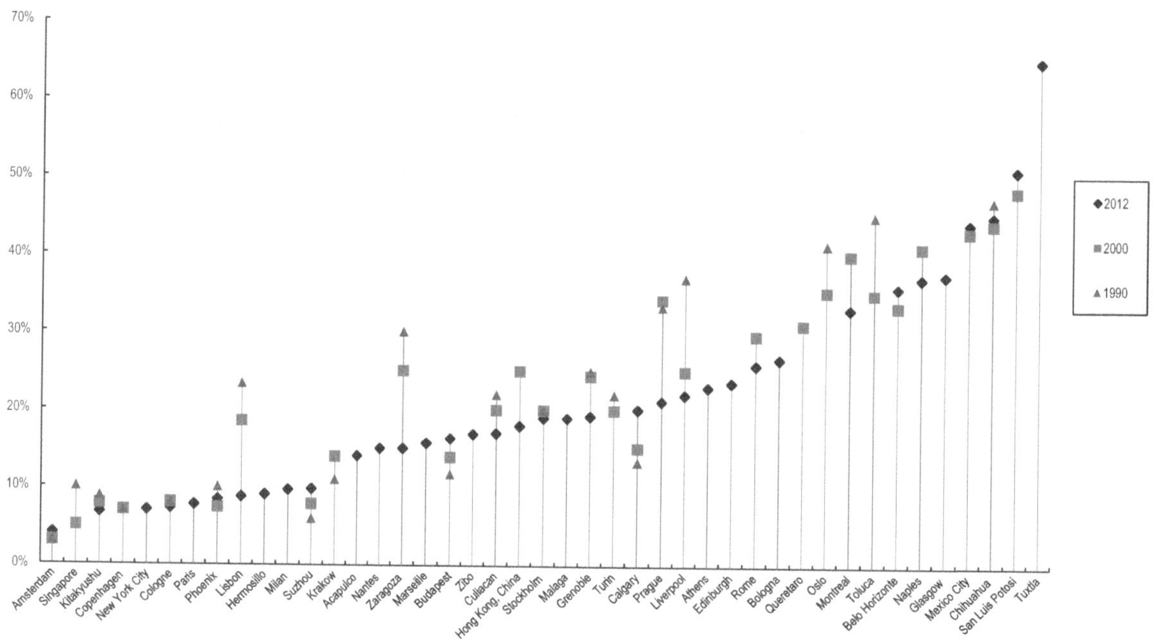

Note: Proportion of water loss as a percentage of net water production (delivered to the distribution system) reported by the surveyed cities. Notes from the surveyed cities: Budapest (data 2013); Liverpool (2012 figure is actual loss for Liverpool. 2000 and 1990 values are based on UU's regional data); Singapore (unaccounted-for-water: PUB monitors the UFW which comprises two components i.e. real losses [leakage] and apparent losses [metering]).

Source: OECD Survey on Water Governance for Future Cities (2014).

Figure 2.4. **Share of water loss by GDP per capita**

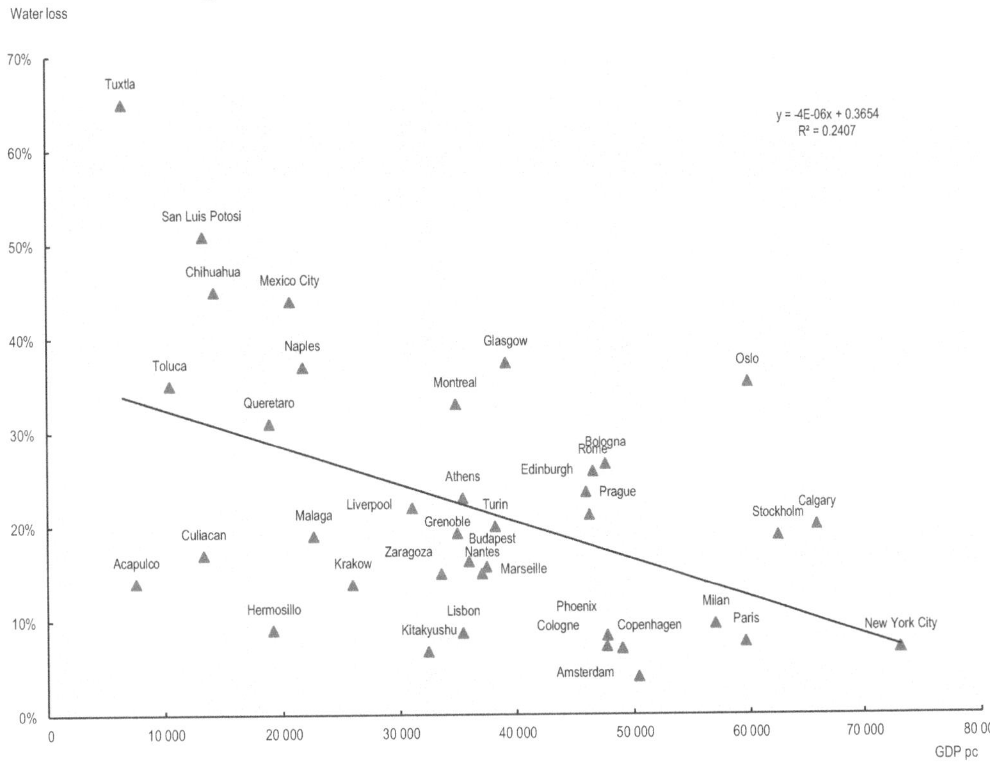

Note: Share of water loss as a percentage of net water production (delivered to the distribution system) reported by OECD surveyed cities (year: 2012). GDP per capita reported in the OECD Metropolitan Areas Database (year: 2010).

Source: OECD Survey on Water Governance for Future Cities (2014).

Infrastructure renewal helps to reduce increasing environmental and operative costs of treatment due to leakages (Box 2.1). In Zaragoza, for example, consistent investments were made to reduce and control water loss such as rehabilitation of the pipeline network and pressure management controls. By 2008, losses from the system were reduced by over 40% compared to 1997, leading to yearly water saving of 20 million m^3 (Philip, 2011). Other cities of the sample have also significantly reduced water losses since the 1990s (Cologne, Grenoble, Kitakyushu, Lisbon, Liverpool, Montreal, Naples, Oslo, Prague, Rome and Stockholm) (Figure 2.4). Improving the information system, flow monitoring and the use of performance indicators related to water losses[1] can also reduce inefficiencies and related environmental and financial costs.

Box 2.1. **Water for the Future in New York City: Fixing leakages**

Since the 1990s, the Department of Environmental Protection (DEP) has been monitoring leaks in a portion of the aqueduct that connect the Rondout Reservoir in Ulster County to the West Branch Reservoir in Putnam County. There are two areas of significant leakage in the Rondout-West Branch Tunnel (RWBT) portion of the Delaware Aqueduct, the Wawarsing and Roseton crossings. Together, they leak approximately 35 million gallons of water per day. In response, DEP plans to construct a bypass tunnel around the leaking areas in Roseton, which would consist of a new tunnel segment to bypass the leaking section and two shafts at each end. This work was started in 2013 and should be completed in 2023. Once the shafts and bypass tunnel are constructed, the aqueduct would be shut down and unwatered. At that time, the leaks in Wawarsing would be repaired and the bypass tunnel would be connected to the existing tunnel. The shut down work would begin in 2022 and take between five to eight months.

Source : Inputs provided by New York City in response to the OECD Survey on Water Governance for Future Cities (2014).

In OECD countries, a total share of 80% of the population is connected to a municipal wastewater treatment plant (OECD, 2015a), but the level of treatment varies across and within countries.

Countries with high levels of tertiary treatment[2] include for example: Austria, Germany, Luxembourg, Netherlands, Spain, Switzerland, United Kingdom (OECD, 2015a). In European cities, this performance is due to the EU Directive concerning urban wastewater treatment, which set standards to be achieved in this area.[3]

In some OECD countries, part of the population is either connected to a wastewater treatment plant with primary treatment only (e.g. Canada, Norway) or to a sewerage network without treatment (e.g. Belgium, Mexico, Portugal, Turkey) (OECD, 2015a).

In OECD countries with relatively low GDP per capita, infrastructure development is ongoing and requires investment in the order of 1% of GDP (OECD, 2015a). According to the results of the OECD Survey on Water Governance for Future Cities, several Mexican cities (e.g. Hermosillo, Veracruz, Toluca and Chihuahua) show lower shares of wastewater treated (Figure 2.5).

The average share of wastewater treatment in the cities surveyed is lower for cities below 1.5 million inhabitants (85%) than for cities with larger populations (98%).

Figure 2.5. **Share of wastewater treated in surveyed cities**

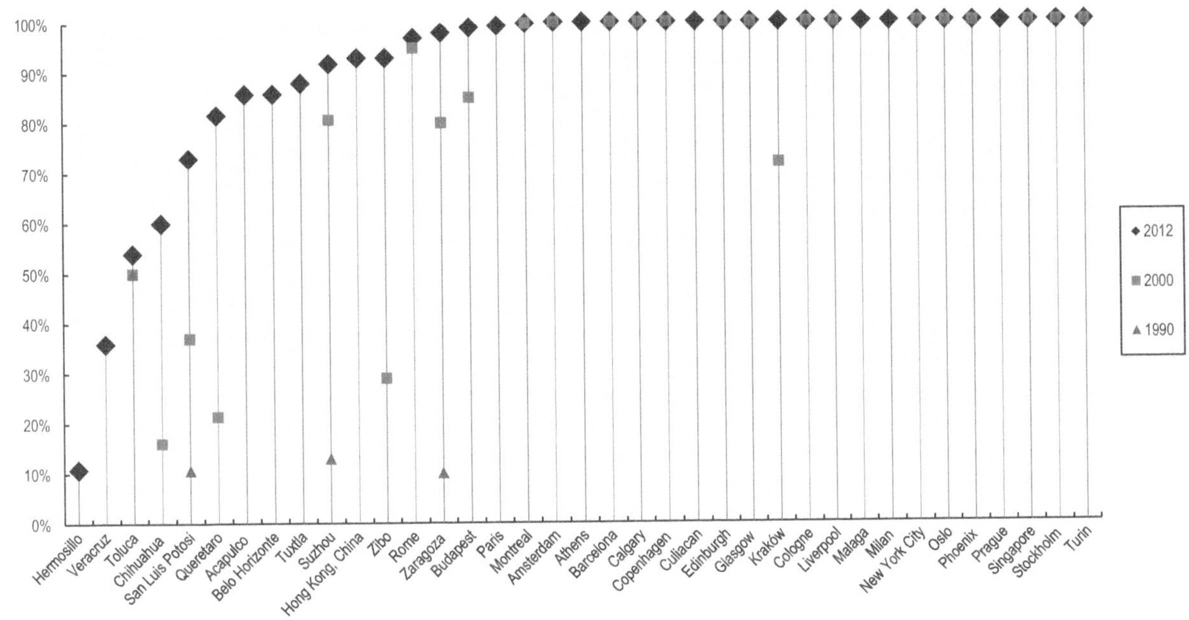

Note: As a percentage of wastewater produced by the city that is collected and treated to at least a basic/primary level. A primary treatment level is physical and/or chemical process involving settlement of suspended solids, or other process in which the BOD5 of the incoming wastewater is reduced by at least 20% before discharge and the total suspended solids are reduced by at least 50% (OECD, 2015a). Barcelona and Queretaro provided data for the metropolitan area.

Source: OECD Survey on Water Governance for Future Cities (2014).

OECD countries have universal coverage for water supply and sanitation services. Overall, 99% of the urban population in OECD countries has access to improved water sources and 98.4% to improved sanitation facilities (World Bank, 2012a). However, this aggregation at national level masks territorial disparities and divides: for example, in France some 200 000 citizens out of 62 million do not receive water from a public network (Smets, 2007). Unconnected fractions of population to water systems are also reported in Galicia, Northern Portugal, Ireland, Southern Italy, Greece, all Eastern European countries and in some Nordic European countries. By 2050, more than 240 million people (mostly in rural areas) worldwide are expected to be without access to an improved water source; almost 1.4 billion people are projected to lack access to basic sanitation (OECD, 2012a). Data for surveyed cities are shown in Figures 2.6 and Figure 2.7.

Figure 2.6. **Share of population with access to safe drinking water**

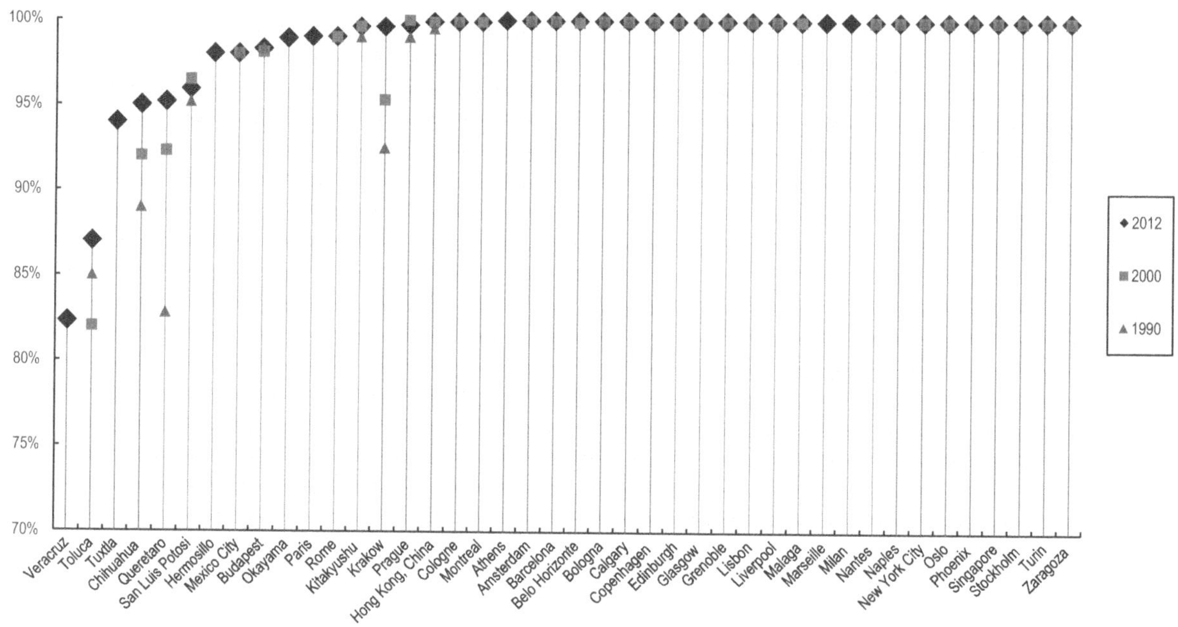

Note: As a percentage of population served by the system. According to the World Health Organization (WHO), access to safe drinking water is "the proportion of people using improved drinking water sources: household connection; public standpipe; borehole; protected dug well; protected spring; rainwater", WHO website, Health through safe drinking water and basic sanitation, http://www.who.int/water_sanitation_health/mdg1/en, (accessed December 2015). Barcelona provided data for the metropolitan area.

Source: OECD Survey on Water Governance for Future Cities (2014).

Figure 2.7. **Share of population with access to sanitation services**

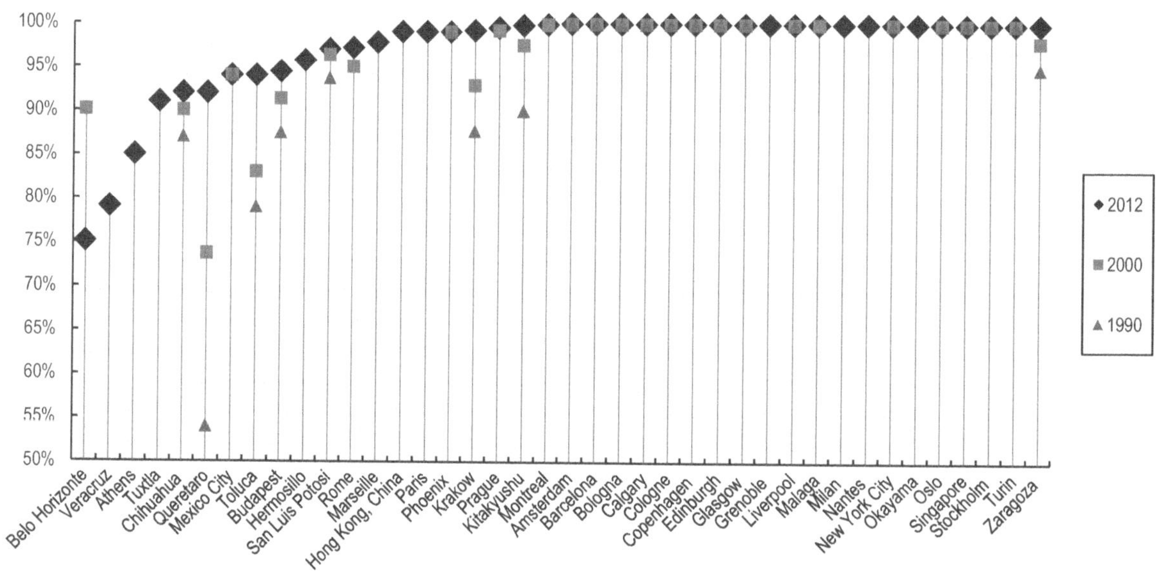

Note: As a percentage of population with direct connections to sewerage, or access to improved on-site sources such as septic tanks and improved latrines. Barcelona provided data for the metropolitan area.

Source: OECD Survey on Water Governance for Future Cities (2014).

Among surveyed cities, the lowest rates in access to sanitation services were reported in Belo Horizonte, where 75% of the urban population is connected to sanitation services (against a national average of 87%; World Bank, 2012b) and in Veracruz (79%). Other surveyed Mexican cities show higher connection rates than the national average (Figures 2.7 and 2.8). Despite progress, Mexico has the lowest rate of connection to public wastewater treatment plants in the OECD; the sanitation gap is expected to reach 4.3 billion cubic metres by 2030 (OECD, 2013a).

Figure 2.8. **Share of population with access to sanitation services in surveyed Mexican cities**

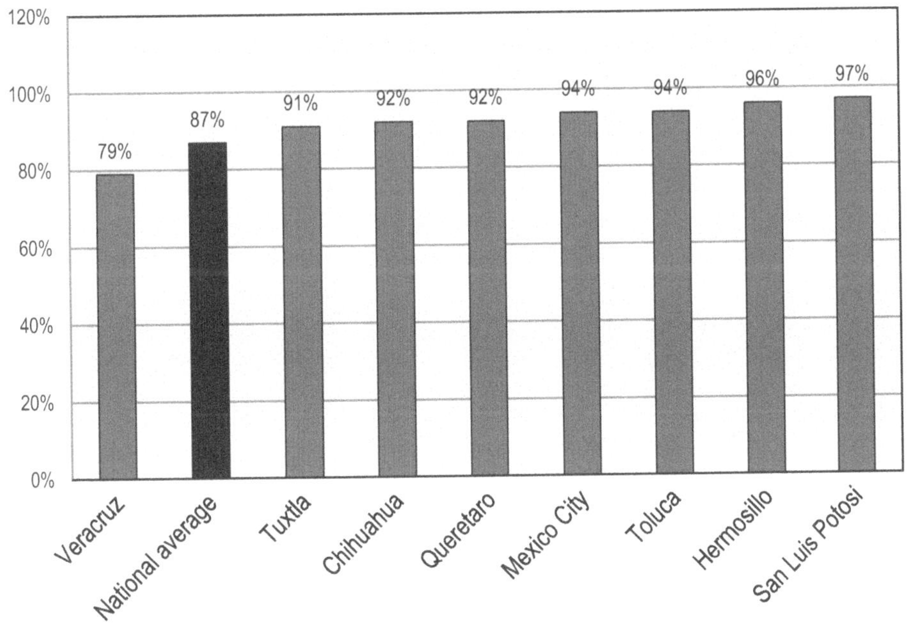

Note: Based on data provided by the surveyed cities. National average data from World Bank indicator "Improved sanitation facilities, urban (% of urban population with access)", year: 2012.

Source: OECD Survey on Water Governance for Future Cities (2014).

The risks of too much, too little, too polluted water

Extreme water-related events (i.e. floods and droughts) are considered challenging factors by 81% of surveyed cities. The number of people at risk from floods will rise from 1.2 billion today, to around 1.6 billion in 2050 (nearly 20% of the world's population) (OECD, 2012a). Floods are one of the most costly and damaging disasters: the severe flooding that hit the city of Copenhagen in 2011 caused about EUR 700 million of damages; hurricane Sandy in New York generated USD 19 billion of economic losses in 2012. The economic value of assets at risk of flood is projected to grow over 340% from 2010, reaching USD 45 trillion. This increase will be much higher for emerging and developing countries as compared to OECD countries (respectively 640%, 440% and 130%) (OECD, 2012a; OECD, 2012c; OECD, 2015b). Projections show that without adaptation measures to floods, there will be losses of USD 1 trillion or more per year (Hallegatte et al., 2013). Recent floods (South of France 2015, Northern Italy 2014, England and Wales, 2012), including from heavy rain, drew attention to the need for greater coherence across water and related areas, in particular for land use; more systematic co-operation across governmental and non-governmental actors (i.e. water companies, local authorities, environmental agencies, residents, urban planners, property developers); greater planning and monitoring capacity at local level for better preparedness; and water sensitive building codes.

Droughts also affect cities. Many cities have implemented measures as result of, or to safeguard against, droughts. Phoenix, Arizona has put in place campaigns to recycle water and reduce water consumption given its location in an arid area and heavy dependence on the Colorado River. In Brazil, a country concentrating 12% of the world's freshwater resources, Rio de Janeiro and São Paulo were hit by the worst drought in 84 years (OECD, 2015c). The São Paulo Water and Waste Management Company (SABESP) provides households with a 30% discount on their water bills if they reduce their water usage by 20% in (Kelman, 2015). Barcelona was also confronted with a very severe drought in 2008, triggering a reflection on Climate Change Adaptation Plans (Guiu, Pouget and Termes, 2015).

Water pollution is a factor requiring greater attention now and in the future for 75% of surveyed cities. Continued investments in wastewater treatment, together with improvements in agriculture, are expected to stabilise and restore surface water and groundwater quality in most OECD countries by 2050 (OECD, 2012a). Outside the OECD, the quality of surface water is expected to deteriorate in the coming decades due to poor wastewater treatment (OECD, 2012a). Proper information systems and urban planning are instrumental to ensure effective water quality management.

The availability of surface and groundwater resources and competition across water uses challenge water managers in the short, medium and long term. Both water demand and supply-side pressures are on the rise. Competition over water allocation represents a challenge for 38% of surveyed cities. Though agriculture remains the largest water user worldwide; by 2050 water demands for manufacturing, electricity and domestic consumption are likely to respectively increase by 400%, 140% and 130% (OECD, 2012a). Overall water availability for irrigation is expected to decline as a consequence of the increase of other uses and sectors (OECD, 2012a). In many countries, water resources are already over used (current levels of abstraction exceed the sustainable level) or over allocated (existing water entitlements – e.g. licenses or permits – to abstract water exceed the sustainable level) (OECD, 2015d). Cities most exposed to "too much" or "too little" water would need to assess and remove governance obstacles to long-term planning in adaptation to climate change, which may also imply actions, such as planting, to create a better micro climate.

Institutional factors

National and international laws and regulations shape water governance at the city level: they represent a major factor for 90% and 67% of cities surveyed, respectively. For example in the Netherlands, the "New Urban Developments and Restructuring" programme within the Delta Programme (2011) sets out objectives and strategies for safety and freshwater supply to be implemented through spatial and urban planning (OECD, 2014a). At international level, the implementation of the EU Water Framework Directive requires building capacities, creating new information systems and a continuous involvement of local actors and citizens for managing water in a sustainable way, co-ordinating public action across levels of government and reducing conflicts at the local level.

A total of 52% of surveyed cities indicated territorial reforms as a factor for changes in urban water governance. Territorial reforms can result in the re-organisation of water services delivery; information sharing across actors initiating new horizontal and vertical interactions; stakeholder engagement; and policy complementarities across different sectors and between cities and surrounding areas. Among OECD countries that have undertaken important territorial reforms (OECD, 2014b), three approaches have been applied with increasing frequency:

1. municipal mergers: to reduce the number of municipalities and increase their scale in terms of geography and population
2. inter-municipal co-operation: to encourage arrangements that allow local jurisdictions to work together for certain common services or investments
3. metropolitan governance: to address the special needs of larger cities and surrounding areas.

In the Netherlands, for example, the number of municipalities has been reduced by more than half following several mergers and re-organisations in the last six decades, and ongoing discussions are targeting a threshold of 100 000 inhabitants per municipality (OECD, 2014a). The ongoing territorial reform in France, which will lead to a merger of the municipalities (36 700, at present) will also have implications on the number of water and sanitation utilities, which are expected to be divided by ten in the coming five years. Other territorial reforms are in place in Australia, Canada, Germany, Iceland, Norway, Switzerland and the United Kingdom (OECD, 2014b). While municipal amalgamation is expected to reach economies of scale; however, in practice transferring responsibilities may require more spending after the merger (OECD, 2014b).

Decentralisation and/or re-allocation of competences were considered a major factor triggering changes in water governance for 46% of cities surveyed. Over the last decades, central governments gradually delegated operational responsibility downwards often to the municipal level (OECD, 2006b). Globally there is a great deal of diversity in institutional and organisational forms for water service delivery. In many developed countries, the city level is the first appropriate territorial level for developing water services and sanitation. More recently, however, to cope with heavy investment costs, local utilities turned towards mergers at upper levels of government for part or all of the tasks to provide and deliver water supply and sanitation services (Barraqué, 2013).

If *upscaling* the management level allows water utilities to achieve economies of scale and to share investment and operation costs of certain equipment, it also offers the opportunity to develop solidarity between territories with different levels of management capacities. Conversely, *downscaling* the level of management offers more autonomous solutions. In both cases, upscaling and downscaling, the governance issue is to ensure *representative* and *participative* democracy at the territorial levels where they operate. This requires both co-ordination across elected representatives in control of public services and the involvement of the broader range of stakeholders. In some cases, a third option is the use of *technology*, without necessarily modifying the institutional set-up for water service management and relations with customers. This was done in Barcelona, which opted for desalination and wastewater recycling to avoid upscaling/downscaling and disputes over water allocation in the region (Barraqué, Isnard and Souriau, 2011).

A number of cities are also subject to liberalisation, privatisation or re-municipalisation trends in the water services sector. Today, liberalisation is associated with the growing role of management and lease contracts, in which the main risks remain with the public authorities, rather than with private actors. This is primarily due to increasing uncertainties, tougher regulation, full cost recovery and long-run investment risks (Menard and Peeroo, 2011). The percentage of the world's population served to some extent by the private sector increased from 5% in 1999 to 14% in 2012 (Pinsent Masons, 2012). On the other hand, the "re-municipalisation" trend has also been seen in several countries. According to the Public Services International Research Unit (PSIRU, 2014), 41 re-municipalisations occurred in high-income countries between 2005 and 2009, a number which doubled between 2010 and 2014. Over the last decade, the early termination or non-renewal of some public-private partnership (PPP) contracts has given way to more open and pragmatic debates and for evolutions in the ways of resorting to the private sector for handling tasks as diverse as management, consulting, auditing, maintenance, sub-contracting and/or joint investment with local authorities (Blanc and Botton, 2012).

Environmental and socio-economic factors

Climate change, urban growth and economic crises are the most prominent exogenous factors that affect water governance in cities (79%, 63% and 58%, respectively). While climate change increases the risk of too much, too little and too polluted water, urban growth represents a threat for the sustainable use of the resource, and economic crises push towards new business models to do more with less.

Bigger and more populated cities are likely to put more pressure on water demand and supply. Global water demand rose by more than double the rate of population growth in the last century (OECD, 2005). An important component of water demand in cities is domestic consumption. The average consumption of domestic water per capita of surveyed cities above 5 million inhabitants is 81.6 m^3 per inhabitant per year; 60.9 m^3 per inhabitant per year in cities below 1.5 million inhabitants; and 66 m^3 per inhabitant per year for the overall sample. The findings show a positive correlation between the population of the urban core of the functional area and per capita domestic water consumption (r=0.3). On average, when population increases by 1%, per capita domestic water consumption is expected to change by 0.12% (Figure 2.9). While the low R^2 value underlines that data are dispersed along the regression line, further tests consisting in the exclusion of the outliers still show a positive correlation. Hence, considering the high heterogeneity of this sample and omitted variable bias, further research would certainly help to shed light on this result. However, the correlation does not necessarily imply causality. A number of components can explain the higher or lower per capita domestic water consumption in cities, such as consumption behaviour, price of water, or meteorological conditions. Typically, the domestic water footprint in the United States is usually higher than in other countries; in some European cities higher water prices tend to raise more awareness on usage.

Figure 2.9. **Total city population and domestic per capita water consumption**

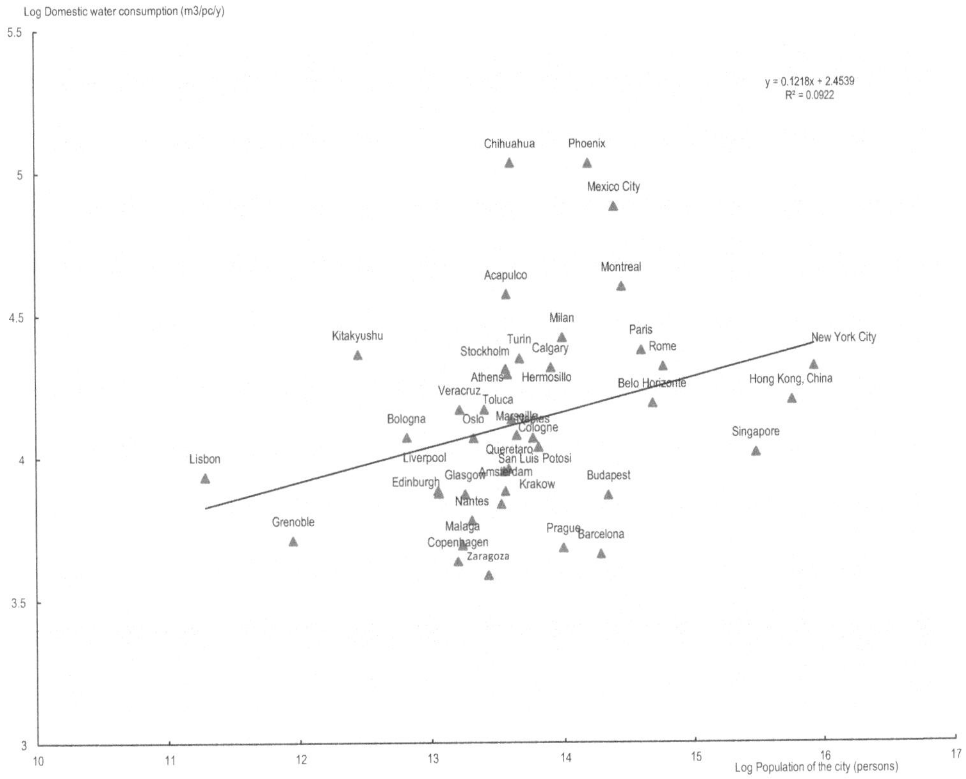

Source: Total city population as reported by surveyed cities from non-OECD countries; National Census (2011; 2010) for cities in OECD countries and Insee (2012), Population en 2012, www.Insee.fr, (accessed December 2015) for the Metropolitan Area of Nantes. Data on domestic water consumption (per capita water consumed by household) was provided by the surveyed cities. Other sources have been used for the following cities: Milan, Turin, Bologna, Naples (Istat, Indicatori sull'acqua per uso domestico per i comuni capoluogo di provincia, http://dati.istat.it/Istat, year: 2011). For the City of Liverpool data are per property and based on regional data.

Overall, there is a trend towards decreasing water consumption per capita over time. This is due to water conservation measures, water loss reduction and enhanced awareness among consumers. In the case of domestic water consumption, customer-targeted water conservation campaigns, systematic monitoring and increased price of water can help reduce the amount of water individually consumed. For instance, a combination of these measures has been successfully employed in Copenhagen, whose inhabitants have consumed a fairly low amount of water compared to other cities of the sample since the 1990s. Copenhagen has completed water-saving campaigns like "Max 100" to raise awareness of citizens on daily water consumption. Likewise, the Calgary's "30-in-30" Plan aims to reduce per capita water demand in the next 30 years (Box 2.2). On the other hand, domestic water consumption per capita has increased in Belo Horizonte, Mexico City, Kitakyushu, Queretaro, Hong Kong, China; and Lisbon among surveyed cities (Figure 2.10). Domestic consumption may also vary within the city's boundaries as in the case of Mexico City where wealthy areas consume up to 600 litres per inhabitant per day of domestic water, while less wealthy areas around 20 litres per inhabitant per day (Universidad Iberoamericana, n.d.).

Figure 2.10. **Per capita domestic water consumption, 1990 -2000-2012**

In m^3 per inhabitant per year

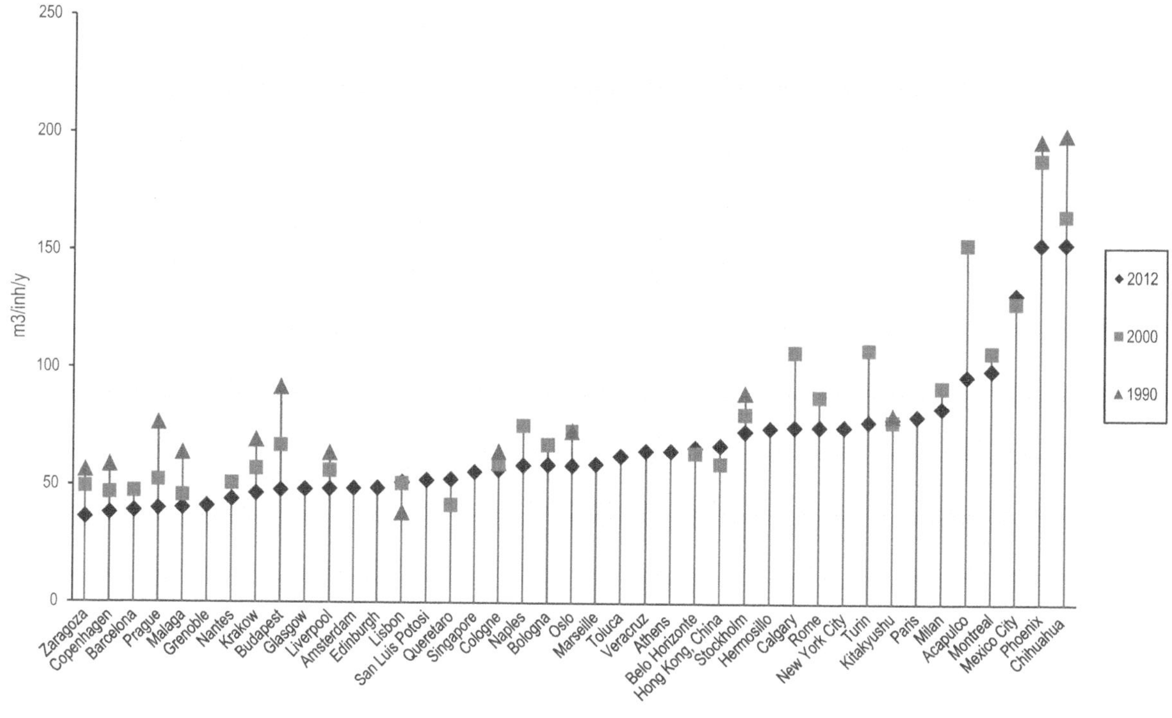

Source: Data provided by surveyed cities in the context of OECD Survey on Water Governance for Future Cities (2014). Nantes provided data for the metropolitan area. Other sources were used for the following cities: Milan, Turin, Bologna, Naples (Istat, Indicatori sull'acqua per uso domestico per i comuni capoluogo di provincial, http://dati.istat.it/Istat, year: 2011) and for Mexico City, Fondo para la Comunicación y la Educación Ambiental (n.d.), Agua de la ciudad de México, www.agua.org.mx, (accessed December 2015); IPN and CONAGUA. (2011), Estudio de manejo de demanda de agua subterránea para el acuífero de la zona metropolitana de la Ciudad de México, www.cuencavalledemexico.com, (accessed December 2015); Iracheta Cenecorta A. (n.d.) Agua y metrópolis: el Valle de México, www.agua.org.mx, (accessed December 2015). For the City of Liverpool data are per property and based on regional data.

The 2008 economic crisis that hit many OECD countries impacted the ability of water utilities and local governments to access credit and to invest in infrastructure operations and maintenance. This was considered a critical or important factor by 58% of surveyed cities, but it might trigger greater search for value for money through increased attention to the efficiency of governance structures (multi-sectoral utilities, local public enterprises, etc.) and innovative sources of finance for renewing ageing infrastructure (property tax, municipal bonds, private equity, long-term institutional investment).

Box 2.2. **Calgary's long-term planning to deal with population growth and increasing water consumption**

The City of Calgary serves a population of more than 1 million people. Calgary experienced unprecedented population growth (approximately 120% growth over the past three decades). Its infrastructure and municipal services must continually expand to keep pace with the demands of a rapidly growing city. As a result, total water use in Calgary is increasing even though the average amount of water used per person is decreasing. The demand for water from the Bow and Elbow rivers also exists beyond the city limits. These rivers are shared by a diverse set of users - other communities, rural populations, industrial users, ranchers, farmers and irrigation districts - all of whom cumulatively exert pressure on the system and add to the complexity of managing this resource. Demand for water from these and other stakeholders is also growing – both upstream and downstream of Calgary. The province is under continued pressure to approve proposals for new licenses and allocations to divert water within the Bow and Elbow watersheds in support of a variety of municipal, recreational, industrial and agricultural activities.

Calgary's "30-in-30" Plan, adopted in 2005, outlines numerous programmes to bring overall water demand down to 350 litres per capita per day in the next 30 years. The goal is to accommodate Calgary's future population growth with the same amount of water removed from the river in 2003. Some of the actions undertaken are: flat rate residences converted to meters; system water loss; water bylaw; customer programmes; and an environmental management system. The City of Calgary identified core strategies based on technological and behavioural changes (see the figure below).

Calgary's water resources efficiency strategies

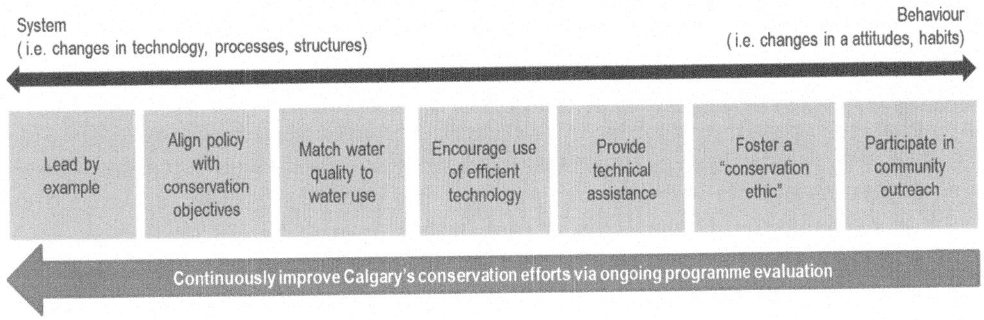

- **Lead by example:** The City aims to set a positive example for customers by ensuring the City is a leader in responsible water use.
- **Align policy with conservation objectives:** A key regulatory component of the City of Calgary's overall water efficiency strategy is to implement universal water metering, which is now completed.
- **Encourage use of water-efficient technologies:** It consists of a series of actions to accelerate wider installation of some water-efficient technologies by providing financial incentives to customers.
- **Provide technical assistance**: The City supports research and provides technical assistance and tools that will help customers achieve measurable water savings.
- **Foster a "conservation ethic" among Calgarians:** The City promotes education campaigns that raise awareness of water efficiency goals, methods and technologies to encourage efficient water use.
- **Engage in community outreach**: The City recognises the value of partnering with stakeholders and the broader community to influence customers' water use behaviours.

Water efficiency programmes are planned for implementation every four years as part of the City of Calgary's four-year budgeting cycle. The City intends to improve water conservation efforts through data collection, monitoring and analysis. Evaluating and adapting their programmes helps to address barriers and meet customer needs.

Source: Inputs provided by the City of Calgary (2007), "Water Efficiency Plan: 30-in-30 by 2033", City of Calgary, Water Resources.

A total of 48% of surveyed cities mentioned that spending decisions and mobilisation of economic instruments to finance water management are affected by fiscal consolidation. Sub-national governments have been consolidating since 2010 and more efforts are required in certain countries to reach debt levels of 2007 or 2010 by the year 2026 (Sutherland, Hoeller and Merola, 2012). Fiscal consolidation has an impact on tax revenues and public budgets as one of the three ultimate sources of water revenues (3Ts), together with tariffs and transfers. It thus holds implications for public investment in water-related infrastructure. User fees in the water sector and other environmentally sensitive goods and services can underpin fiscal consolidation by managing and containing demand (OECD, 2013b). Other sources of public finance are also being contemplated, e.g. property taxes, which can help ensure that those who generate future liabilities (e.g. property developers) bear some of the related costs (see Chapter 4, the financing section).

Summary of factors that affect different types of surveyed cities

While it is possible to identify common trends across surveyed cities, the impacts of certain factors differ across cities according to their characteristics (e.g. size, population, spatial organisation, etc.) and institutional organisation. Table 2.1 provides a scoreboard illustrating the impact of several factors on urban water governance across the categories of cities by size, density, demographic dynamics, and metropolitan governance arrangements described in Chapter 1[4].

Results from the sample show that the impact of factors is more or less acute according to the size of a city:

- The six surveyed cities with more than 5 million inhabitants listed the following factors as critical or important: ageing, obsolete infrastructure, extreme events, water pollution, climate change and urban growth. Their distinctive feature compared with smaller cities is therefore the combination of multiple factors and related challenges.
- The 15 surveyed cities with a population between 1.5 and 5 million inhabitants ranked national laws and regulations, ageing infrastructure and climate change as the first three critical and important factors that affect water management.
- The 27 surveyed cities with less than 1.5 million inhabitants identified ageing infrastructure and national laws and regulation as major factors. Compared to the other two categories, this latter group shows higher concerns for emergency-driven management and territorial reform.

Reading the factors through the lens of the spatial patterns reveals similarities and differences in the prominent factors that influence water governance in cities:

- Both relatively high (29 cities) and low (19 cities) densely populated cities are predominately concerned with ageing infrastructure, national laws and regulation and extreme events.
- Compared to the lower density conurbations, those with high level of population density (29 cities) show greater concern for liberalisation trends and implementation of the human right to water and sanitation.

- Less densely populated cities (19 cities) were more concerned with fiscal consolidation, decentralisation/ re-allocation of competences and competition of water allocation.

Table 2.1. **Summary of factors that affect different types of surveyed cities**

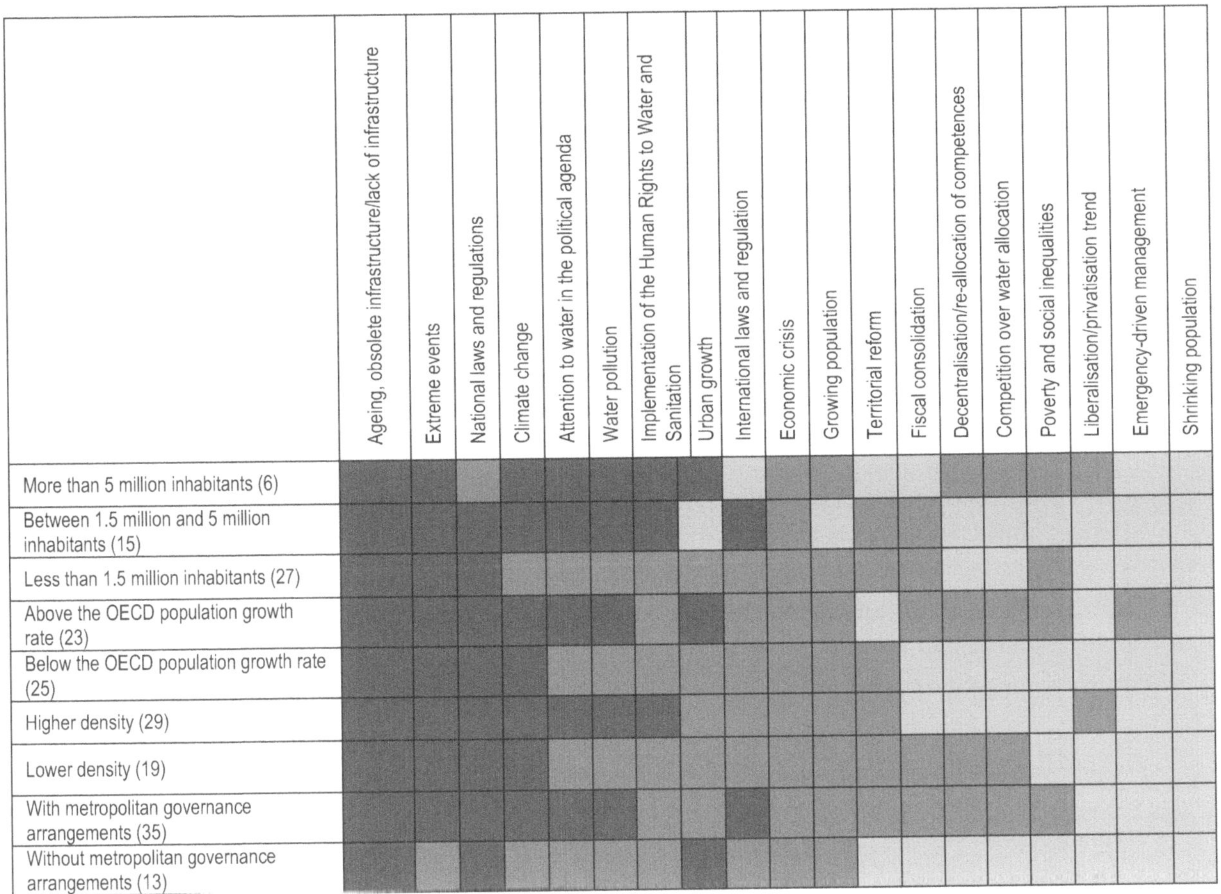

Legend: The colour system indicates the percentage of responses for each identified group of cities within the four macro-categories (size, spatial patterns, demographic dynamics, metropolitan governance arrangements). Light blue: = below 45%; Blue = between 45% and 70%; Dark blue = above 70%.

Looking at demographic dynamics, some observations can also be made:

- In the 25 cities growing below the population average, the factors likely to trigger the most changes in water governance are national laws and regulation, ageing infrastructure, extreme events. A total of 32% of surveyed cities also reported concerns about shrinking population.
- In the 23 cities with population growth above the OECD average, extreme events, national laws, ageing infrastructure, climate change and water pollution came first. Urban growth and competition over water allocation were also identified as distinctive factors.

The presence of metropolitan governance arrangements also makes a difference:

- Cities with metropolitan governance arrangements (35 cities) ranked ageing infrastructure, national laws and extreme events as the most prominent factors.

- Cities without metropolitan governance arrangements (13 cities) ranked ageing infrastructure first, national laws and extreme events second and urban growth third.
- Cities with metropolitan arrangements (even if not necessarily set with the aim of dealing with water) indicated greater concern for all the factors, with the exception of urban growth and liberalisation trends, compared to the responses provided by the group of cities where metropolitan governance arrangements are not in place.

Notes

1. For example, IWA (Alegre et al., 2006) suggests a number of non-revenue water and water loss performance indicators (e.g. non-revenue water by volume: percentage of the system input volume that corresponds to nonrevenue water; non-revenue water by costs: percentage of the system input volume that corresponds to non-revenue water, water losses: m3/service connection/year).
2. Tertiary treatment: treatment of nitrogen and/or phosphorous and/or any other pollutant affecting the quality or a specific use of water (microbiological pollution, colour, etc.), (OECD, 2015a).
3. Directive 91/271/EEC, amended by Directive 98/15/EC.
4. The clustering is based on data at metropolitan level (OECD 2012d), however challenges refers to the main city of the metropolitan area. In the case of Barcelona and Nantes responses were provided for the metropolitan area.

Bibliography

Alegre, H. et al. (2006), *Performance Indicators for Water Supply Services*, Second edition, Manuals of Best Practice, IWA Publishing.

Barraqué, B. (2013), "New modes of water and sanitation services management and emerging business model", working paper developed under Output area 2.3.2 Climate Change and Natural Resource Management of the WPBWE Work Programme for 2011.

Barraque, B., L. Isnard and J. Souriau (2011), "Evolutions possibles de la gouvernance interne et externe des services d'eau en France à la lumière des évolutions européennes", LIVRABLE 5.2, ANR Villes Durables 2008.

Blanc, A. and S. Botton (dir.) (2012), "Water services and the private sector in developing countries: Comparative perceptions and discussion dynamics", AFD and PPIAF, Paris.

City of Calgary (2007), "Water Efficiency Plan: 30-in-30 by 2033", City of Calgary, Water Resources.

European Commission Directive 98/15/EC of 27 February 1998 amending Council Directive 91/271/EEC with respect to certain requirements established in Annex I, Official Journal of the European Communities, L 67/29.

European Council Directive of 21 May 1991 concerning urban waste water treatment (91/271 /EEC), Official Journal of the European Communities, No L 135/40.

European Commission (2013), "Resource and economic efficiency of water distribution networks in the EU", final report by ERM and Solventa, available at: http://ec.europa.eu/environment/water/quantity/pdf/Final%20REE%20Report%20Oct%202013.pdf.

Farley, M. (2001), *Leakage Management and Control: A Best Practice Training Manual*, World Health Organization, Geneva, available at: http://whqlibdoc.who.int/hq/2001/WHO_SDE_WSH_01.1_eng.pdf.

Fondo para la Comunicación y la Educación Ambiental (n.d.), Agua de la ciudad de México, www.agua.org.mx, (accessed December 2015).

Guiu, R., L. Pouget and M. Termes (2015), "Selecting an efficient adaptation level to uncertain water scarcity by coupling hydrological modelling and economic valuation", *Water Economics and Policy*, Vol. 1, No. 3.

Hallegatte, S. et al. 2013), "Future flood losses in major coastal cities", *Nature Climate Change* 3, pp. 802-806, www.nature.com/nclimate/journal/v3/n9/full/nclimate1979.html, (accessed November 2015).

Hammer, S. et al. (2011), "Cities and green growth: A conceptual framework", *OECD Regional Development Working Papers* 2011/08, OECD Publishing, Paris, http://dx.doi.org/10.1787/5kg0tflmzx34-en.

Insee (2012), Population en 2012, www.insee.fr, (accessed November 2015).

IPN and CONAGUA (2011), Estudio de manejo de demanda de agua subterránea para el acuífero de la zona metropolitana de la Ciudad de México, www.cuencavalledemexico.com, (accessed December 2015).

Iracheta Cenecorta A. (n.d.), Agua y metrópolis: el Valle de México, www.agua.org.mx, (accessed December 2015).

Istat (2011), Indicatori sull'acqua per uso domestico per i comuni capoluogo di provincia, http://dati.istat.it, (accessed September 2015).

Kelman, J. (2015), "Water supply to the two largest Brazilian metropolitan regions", *Aquatic Procedia,* Vol. 5, pp. 13-21, World Water Week, 31 August-5 September 2014, Stockholm, Sweden.

Ménard, C. and A. Peeroo (2011), "Liberalization in the water sector: Three leading models", in Kunneke, Rolf and Matthias Finger, *International Handbook of Network Industries: The Liberalization of Infrastructures*, Edward Elgar, pp. 310-327.

McKinsey and Co. (2013), "Infrastructure productivity: How to save $1 trillion a year", January, in WWC and OECD (2015), "Water: Fit to finance?", Report of the High Level Panel on Financing Infrastructure for a Water-Secure World, April.

OECD (2015a), *Environment at a Glance 2015: OECD Indicators*, OECD Publishing, Paris, http://dx.doi.org/10.1787/9789264235199-en.

OECD (2015b), *Water and Cities. Ensuring Sustainable* Futures, OECD Publishing, Paris, http://dx.doi.org/10.1787/9789264230149-en.

OECD (2015c), *Water Resources Governance in Brazil*, OECD Publishing, Paris, http://dx.doi.org/10.1787/9789264238121-en

OECD (2015d), *Water Resources Allocation: Sharing Risks and Opportunities*, OECD Studies on Water, OECD Publishing, Paris, http://dx.doi.org/10.1787/9789264229631-en.

OECD (2014a), *Water Governance in the Netherlands: Fit for the Future?,* OECD Studies on Water, OECD Publishing, Paris, http://dx.doi.org/10.1787/9789264102637-en.

OECD (2014b), *OECD Regional Outlook 2014: Regions and Cities: Where Policies and People Meet*, OECD Publishing, Paris, http://dx.doi.org/10.1787/9789264201415-en.

OECD (2013a), *OECD Environmental Performance Reviews: Mexico 2013*, OECD Publishing, Paris, http://dx.doi.org/10.1787/9789264180109-en.

OECD (2013b), "Fiscal consolidation across levels of government", in OECD, *Fiscal Federalism 2014: Making Decentralisation Work*, OECD Publishing, Paris, http://dx.doi.org/10.1787/9789264204577-3-en

OECD (2012a), *OECD Environmental Outlook to 2050: The Consequences of Inaction*, OECD Publishing, Paris, http://dx.doi.org/10.1787/9789264122246-en.

OECD (2012b), "Review of the OECD Environmental Strategy for the first decade of the 21st century", OECD, Paris, available at: www.oecd.org/env/50032165.pdf.

OECD (2012c), *Meeting the Water Reform Challenge*, OECD Studies on Water, OECD Publishing, Paris, http://dx.doi.org/10.1787/9789264170001-en.

OECD (2012d), "Metropolitan areas", *OECD Regional Statistics* (database), http://dx.doi.org/10.1787/data-00531-en, (accessed October 2015).

OECD (2011), *Meeting the Challenge of Financing Water and Sanitation: Tools and Approaches*, OECD Studies on Water, OECD Publishing, Paris, http://dx.doi.org/10.1787/9789264120525-en.

OECD (2007), *Infrastructure to 2030 (Vol.2): Mapping Policy for Electricity, Water and Transport*, OECD Publishing, Paris, http://dx.doi.org/10.1787/9789264031326-en.

OECD (2006a), "Water: The experience in OECD countries", background documentation to the meeting of the Fourth World Water Forum (Mexico, March 2006), OECD, Paris, available at: www.oecd.org/env/resources/36225960.pdf.

OECD (2006b), *Infrastructure to 2030: Telecom, Land Transport, Water and Electricity*, OECD Publishing, Paris, http://dx.doi.org/10.1787/9789264023994-en.

OECD (2005), *OECD Factbook 2005: Economic, Environmental and Social Statistics*, OECD Publishing, Paris, http://dx.doi.org/10.1787/factbook-2005-en

Philip R. (2011), "Reducing water demand and establishing a water saving culture in the City of Zaragoza", Case study: Zaragoza, Spain, *SWITCH Training Kit*, available at:

www.switchtraining.eu/fileadmin/template/projects/switch_training/files/Case_studies/Zaragoza_Case_study_preview.pdf.

Pinsent Masons (2012), *Pinsent Masons Water Yearbook 2012–2013*, Pinsent Masons LLP, London.

Public Services International Research Unit (PSIRU), Multinational Observatory and Transnational Institute (TNI) (2014), "Here to stay: Water remunicipalisation as a global trend", available at: www.tni.org/files/download/heretostay-en.pdf.

Smets, H. (2007), "Implementing the right to water in France", Water Academy France, available at: www.ielrc.org/activities/workshop_0704/content/d0723.pdf.

Sutherland, D., P. Hoeller and R. Merola (2012), "Fiscal consolidation: How much, how fast and by what means?", *OECD Economic Policy Papers*, No. 1, http://dx.doi.org/10.1787/5k9bj10bz60t-en.

Universidad Iberoamericana (n.d.), "La problemática del agua en México y el mundo", available at: www.uia.mx/web/files/la_problematica_del_agua%20.pdf.

WHO website, Health through safe drinking water and basic sanitation, http://www.who.int/water_sanitation_health/mdg1/en, (accessed December 2015).

World Bank (2012a), *World Development Indicators* (database), http://data.worldbank.org/, (accessed September 2015).

World Bank (2012b), *Improved sanitation facilities, urban (% of urban population with access)* (database), http://data.worldbank.org, (accessed September 2015).

Chapter 3

Mapping who does what in urban water governance

Understanding who does what, at which level and how in water policy design and implementation is a first step to identify potential mismatches, overlaps, grey areas and to suggest ways forward for better co-ordination across multiple scales, authorities and policy domains. This chapter provides an institutional mapping of key water management functions in surveyed cities, by analysing the allocation of roles and responsibilities of central governments, local governments, service providers and other actors at the sub-national level for drinking water, sewage collection, wastewater treatment, drainage and water security.

Key messages

This chapter argues that urban water governance is a shared responsibility across levels of government and a broad range of stakeholders and that co-ordination among these is needed to minimise fragmentation and overlaps.

Water management takes place at several scales depending on the service required (protection against floods or droughts, water supply, sanitation, drainage, etc.) and it is scattered across a number of actors. This reflects the fragmentation of the sector, as well as its complexity.

Water is mostly managed locally but it has impacts at the national and global level. Therefore, responsibilities for policy making, information/monitoring/evaluation, regulation and financing are often shared across levels of government.

Even in highly decentralised contexts, national governments have a role in water policy; therefore cities need to work with upper levels of government, and vice versa, when designing and implementing water policies.

Central governments generally play a higher role for water security, water supply and sanitation than for sewage collection and drainage, which remain prominently responsibilities of local levels. Local governments hold a more operative role. The second tier of government tends to play an important role in information/monitoring and evaluation.

When roles and responsibilities are unclear or overlapping they hinder effective water policy design and implementation (e.g. delays, high transaction costs, asymmetry of information, etc.). Mapping *who* does *what* is the first step to clarify the allocation of roles and responsibilities of actors at different levels of government and across water management functions to foster effective, efficient and inclusive policies.

Central governments

Among cities surveyed, central governments often play a role in urban water management. This role is somewhat stronger in unitary countries as opposed to federal or quasi-federal ones that have largely decentralised key responsibilities to lower levels of government.

When analysing responsibilities by *governance function*, a key observation is that when central governments play a role in urban water management, it is mostly through the lens of policy making and regulation (regardless the water management function), rather than information/evaluation and financing (Figure 3.1).

Figure 3.1. **Roles and responsibilities of central governments in water management**

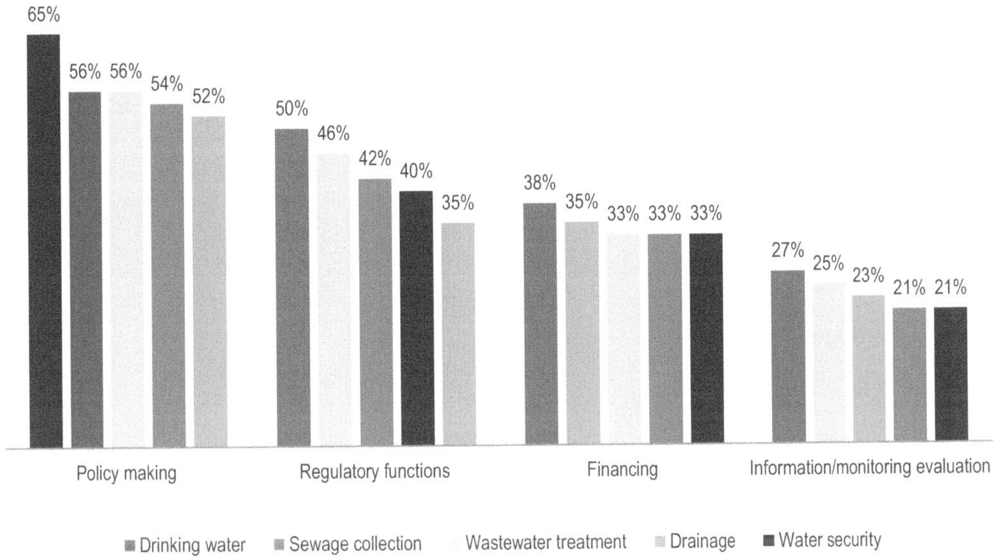

Note: Percentage calculated on the number of responses of the 48 cities surveyed indicating "central government" as has having a role in the investigated water functions.

Source: OECD Survey on Water Governance for Future Cities (2014).

When zooming further by *water management function*, central governments tend to play an important role in policy making/implementation for water security[1] (65%), and they are also heavily involved in the regulation of drinking water and wastewater treatment (Box 3.1).

The roles and responsibilities of central governments in water-related tasks vary by governance function and sub-sector (Figure 3.1):

- More than half of surveyed cities indicated that central governments contribute to policy making in drinking water supply (56%), wastewater treatment (56%) and sewage collection (54%). The lower percentage for drainage indicates that this is mainly a responsibility of sub-national governments.
- Central governments play a prominent role in terms of regulation for drinking water supply (50%) and wastewater treatment (46%). For water supply, countries may have different institutional organisation. Globally, however, a trend towards establishing dedicated economic water regulatory bodies can be observed, where the regulators deal, amongst others, with tariff regulation and performance monitoring, e.g. Italy with *Autorità per l'energia elettrica il gas ed il sistema idrico* (AEEGSI, National Authority of Electricity, Gas and Water Services), Portugal with *Entidade Reguladora dos Serviços de Águas e Resíduos* (ERSAR, Water and Waste Services Regulation Authority), England and Wales with OFWAT (OECD, 2015a). This trend generally accompanies a reform of the water industry towards the consolidation of the water service provision around fewer but bigger providers. Such regulators work with national and sub-national actors as well as various public agencies.

Box 3.1. **Water management functions and governance functions**

The OECD Survey on Water Governance for Future Cities investigated roles and responsibilities of different levels of governments for each of the following water management functions:

- drinking water supply
- sewage collection
- wastewater treatment
- drainage
- water security.

Roles and responsibilities targeted in the institutional mapping are:

- policy design, e.g. strategy, priority setting and planning (including infrastructure)
- implementation of policies at the territorial level
- regulatory functions, e.g. allocation of uses, quality standards, economic regulation
- information, monitoring and evaluation
- financing, e.g. authorities involved in water policy budgets.

Key water management functions, roles and responsibilities

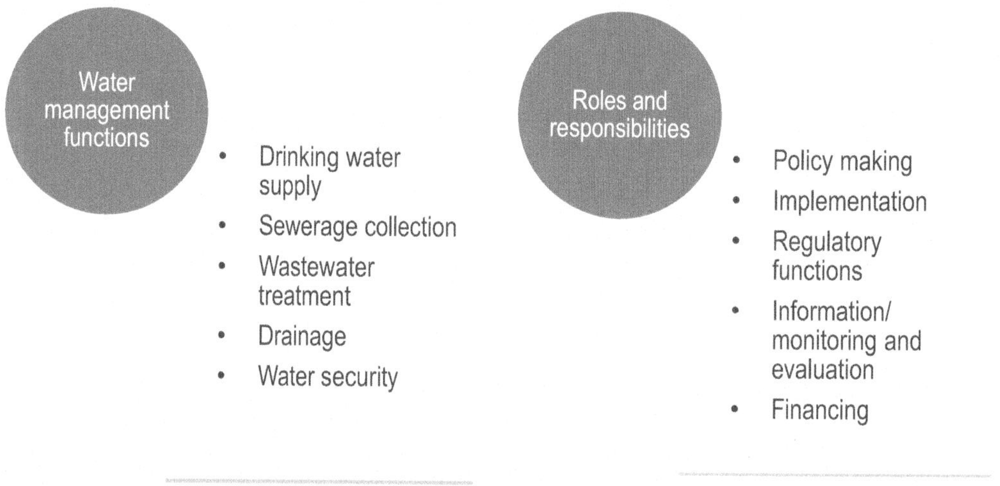

Source: Author's elaboration.

- A fewer number of cities surveyed indicated that central governments contribute to financing for drinking water supply (38%), drainage (35%) and to a lesser extent sewage collection (33%).
- Central governments tend to play a less predominant role overall in information, monitoring and evaluation: below 30% regardless of water management functions. This is primarily due to the privileged position of sub-national governments in accessing related information at the local level.

Based on the comparison of roles and responsibilities at the central and sub-national level across the cities surveyed, three types of models across levels of government can be proposed (Figure 3.2). The first portrays a dominant role of central actors; the second shows a dominant role of sub-national actors, and the third reflects a shared responsibility among central and sub-national level for policy making.

Sub-national governments

Decentralisation of water policies across OECD countries increased the role of sub-national governments in water management (OECD, 2011a). There is no unique model for managing water in cities but a range of options reflecting a diversity of situations within and across countries. However, some observations can be drawn from the institutional mapping of the surveyed sample in terms of the role of local governments (Figure 3.3), state/provincial/regional governments (Figure 3.4) and inter-municipal or metropolitan bodies (Figure 3.5).

Figure 3.2. **Typology of allocation of roles and responsibilities for urban water policy making**

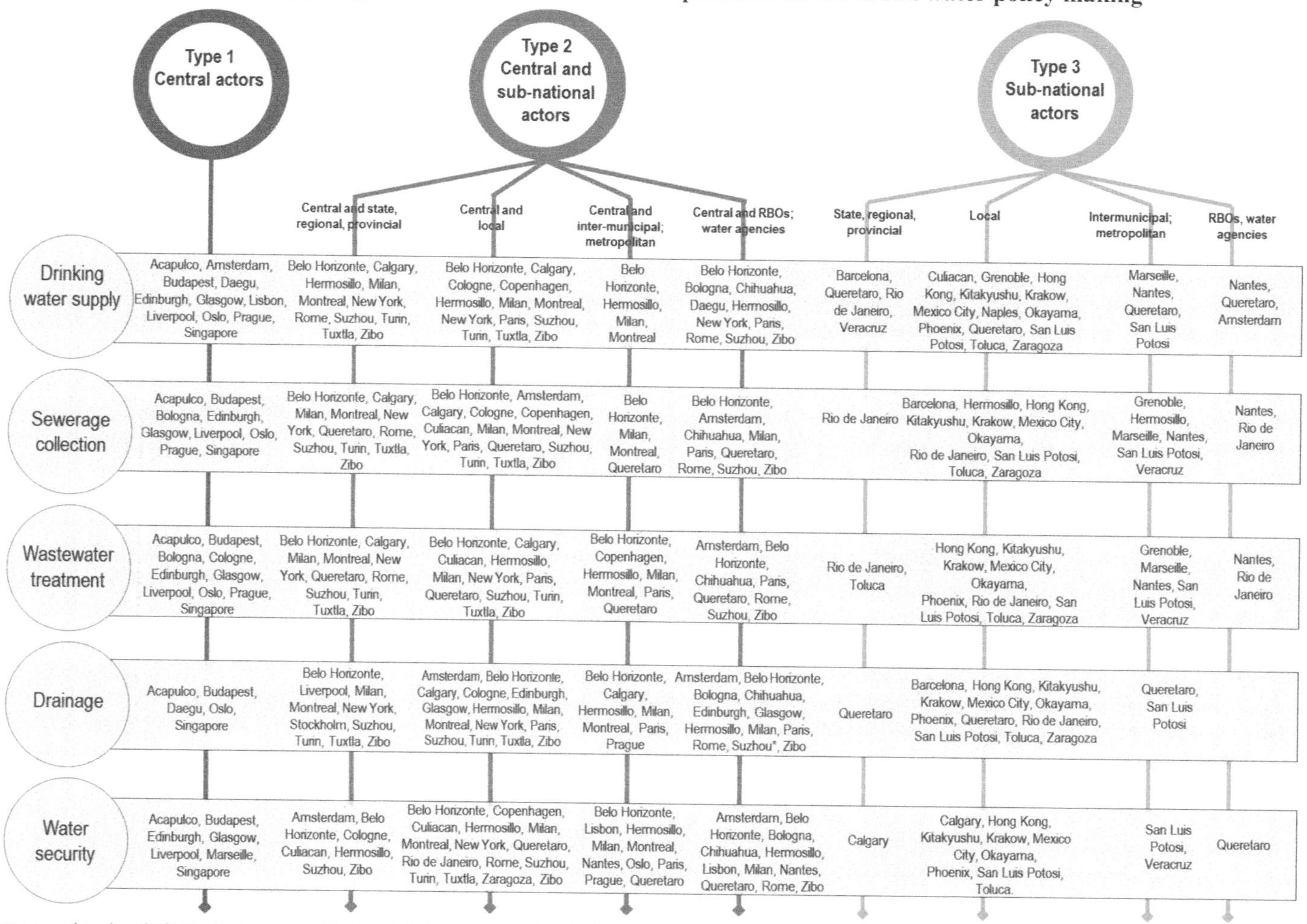

Note: Type 1: Cities that reported a role only for "central government"; Type 2: Cities that reported a role for several categories of sub-national actors; Type 3: Cities that reported a role for both central and sub-national actors.
Source: OECD Survey on Water Governance for Future Cities (2014).

Figure 3.3. **Roles and responsibilities of local governments in water management**

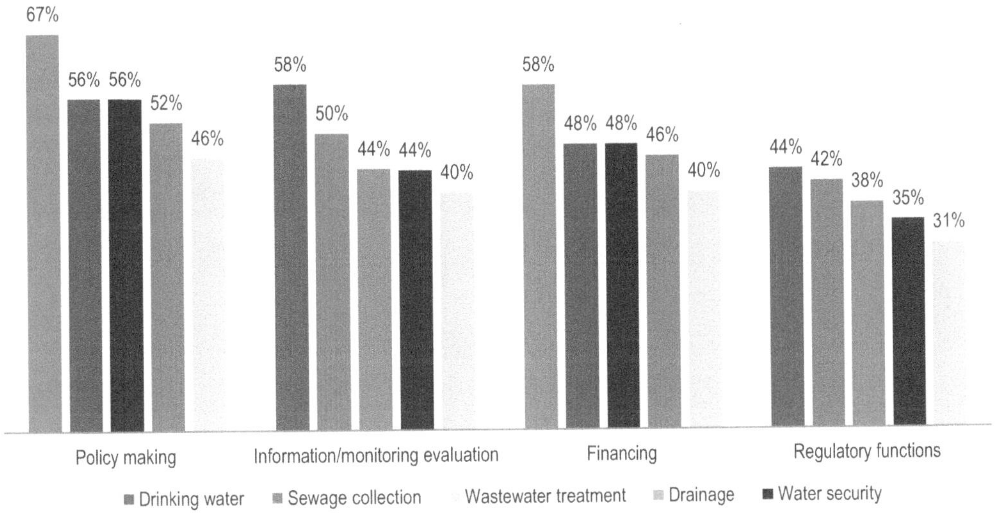

Note: Percentage calculated on the number of responses of the 48 cities surveyed that indicated "local government" as having a role in the investigated water functions.

Source: OECD Survey on Water Governance for Future Cities (2014).

Figure 3.4. **Roles and responsibilities of state, regional and provincial governments in water management**

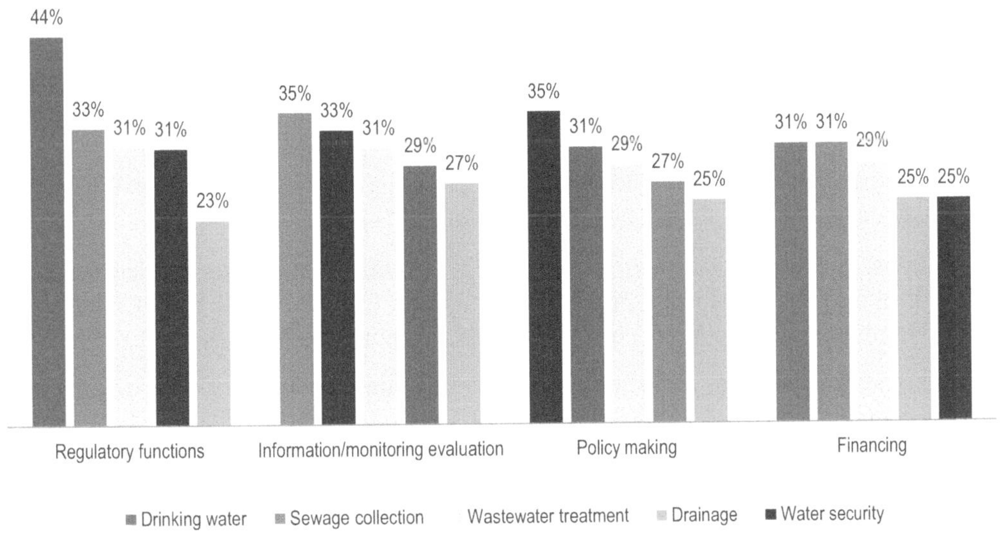

Note: Percentage calculated on the number of responses of the 48 cities surveyed that indicated "state, regional and provincial government" as having a role in the investigated water functions.

Source: OECD Survey on Water Governance for Future Cities (2014).

Figure 3.5. **Roles and responsibilities of inter-municipal, supra-municipal and metropolitan bodies in water management**

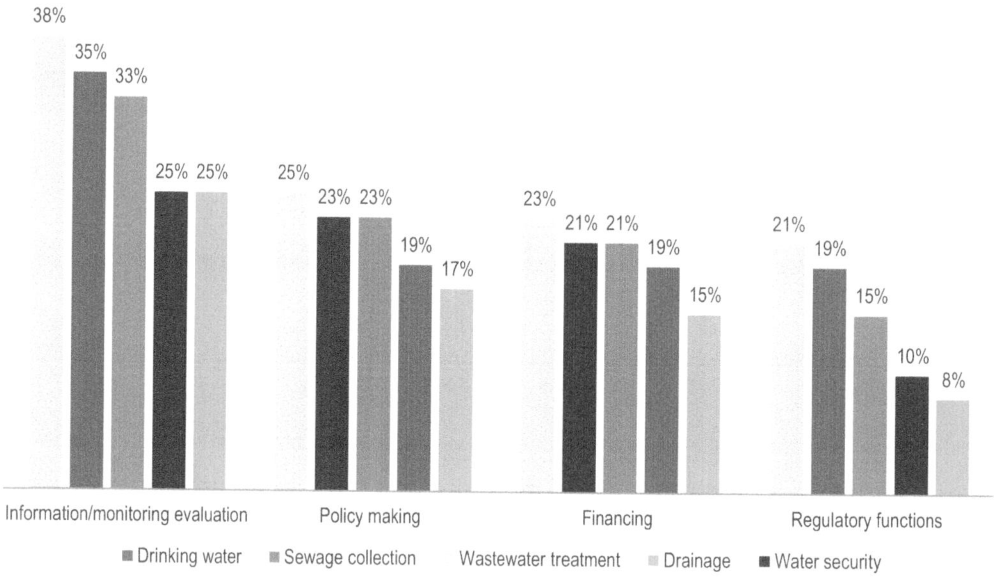

Note: Percentage calculated based on the number of responses of the 48 cities surveyed that indicated "inter-municipal, supra-municipal, metropolitan body" as having a role in listed water management functions.

Source: OECD Survey on Water Governance for Future Cities (2014).

In most cities surveyed, local governments (municipalities) are the primary sub-national authorities in charge of designing and/or implementing policies for drinking water supply and wastewater services, followed by state, regional, provincial governments, and inter-municipal, supra-municipal, metropolitan bodies. In most cases, drinking water and drainage are also a shared responsibility across multiple sub-national authorities.

In terms of policy making, the role of local governments compared to other sub-national actors is definitely predominant for drainage (67%), drinking water supply (56%), water security (56%) and sewage collection (52%).

- Local governments share responsibilities with the central government for drinking water supply, in Belo Horizonte, Calgary, Cologne, Copenhagen, Hermosillo, Milan, New York City, Paris, Suzhou, Turin, Tuxtla and Zibo.[2]
- Local governments carry out regulatory functions in 44% of cases for drinking water supply and 42% for sewage collection.
- With no exception across water functions, the majority of cities indicated that local governments are the main actors providing information and carrying out monitoring and evaluation. This is particularly true for drinking water supply (58%).
- Compared to the responses attributed to central governments and to other sub-national governments, the highest share of responses was attributed to local governments for financing related to drainage (58%); water security and drinking water (48%); sewage collection (46%); and wastewater treatment (40%).

The role of the second tier of government (state, regional and provincial) varies across water management functions.

- The highest share relates to regulation (44%), in the case of drinking water supply and sewage collection (33%); information, monitoring and evaluation in the case of sewage collection (35%) and water security (33%); policy making in the case of water security (35%).
- State, regional and provincial governments are in charge of financing in a relatively lower number of cases compared to central and local governments.
- Where they exist, and with no exception across water management functions, inter-municipal, supra-municipal or metropolitan bodies play a role in information/monitoring and evaluation (Figure 3.5).
- When they contribute to policy making, their role is never exclusive, but they share responsibilities with other levels of governments, including the central government.
- For all water management functions, inter-municipal, supra-municipal or metropolitan bodies carry out regulatory functions and have financing prerogatives in less than 25% of the sample.

Other actors at the sub-national level

Several countries have set up river basin organisations (RBOs) to manage water within integrated basin governance systems (Figure 3.6). River basin organisations are official organisations set up by political authorities or in response to stakeholders' demands or legal requirements. As administrative and hydrologic perimeters do not coincide in most cases, a diversity of situations can be observed in terms of how many RBOs operate within a city's perimeter (from one to more than three; see Figure 3.6). River basin organisations carry out different water-related tasks such as monitoring (85%), data collection (81%), as well as co-ordination, planning and stakeholder engagement (above 60%). In 28% of surveyed cities, RBOs collect taxes or levies (Figure 3.7). Cities usually take part in the governance of the RBOs in several ways: sitting in the directive committees (Belo Horizonte, Calgary, Grenoble, Lisbon, Nantes, New York City, Oslo, Prague, Rio de Janeiro, Rome, Toluca, Zaragoza); voting or taking decisions (Culiacan); participating in meetings (Athens, Mexico City, Naples); being consulted for river basin plans (Edinburgh, Glasgow); and/or advising the committee (Malaga, Queretaro).

Figure 3.6. **Number of river basin organisations within surveyed cities' administrative perimeters**

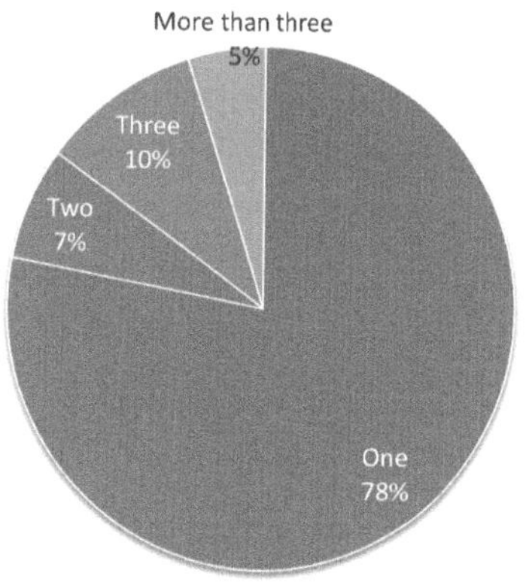

Note: Based on 41 cities that provided the number of RBOs operating within their cities' administrative perimeters.

Source: OECD Survey on Water Governance for Future Cities (2014).

- **One RBO:** Acapulco, Amsterdam, Athens, Barcelona, Belo Horizonte, Bologna, Budapest, Chihuahua, Copenhagen, Culiacan, Edinburgh, Glasgow, Grenoble, Hermosillo, Lisbon, Liverpool, Malaga, Marseille, Mexico City, Milan, Naples, Paris, Phoenix, Roma, San Luis Potosi, Suzhou, Toluca, Turin, Tuxtla, Veracruz, Zaragoza, Zibo.
- **Two RBOs:** Queretaro (La Cuenca Lerma-Santiago RH 12 and La Cuenca del Panuco RH 26); New York City (Delaware River Basin Commission, Decree Party Principles); Prague (Elbe River Basin Authority and Vltavy River Basin Authority).
- **Three RBOs**: Calgary (Bow River Basin Council, Elbow River Watershed Partnership, Calgary Nose Creek Watershed Partnership); Oslo (Oslo; Leira/Nitelva; Indre Oslofjord Vest); Rio de Janeiro (RBO Paraiba do Sul; CBH Rio Guandu; CBH Baia de Guanabara); Stockholm (Svealand Coastal Conservation Association; Tyresan Water Conservation Association, Lake Malaren Water Conservation Association).
- **Five RBOs**: Nantes (Agence de l'Eau Loire Bretagne, Groupement d'Intérêt Public Loire Estuaire [GIP], Loire Estuaire, EDENN [Entente pour le Développement de l'Erdre Navigable et Naturel], EPTB [Établissement Public Territorial de Bassin] de la Sèvre Nantaise, SEVRAVAL [Syndicat Sèvre Aval]); Cologne (IKSR, IAWR, ARW, Erftverband, Wupperverband).

Figure 3.7. **Key functions of the river basin organisations of surveyed cities**

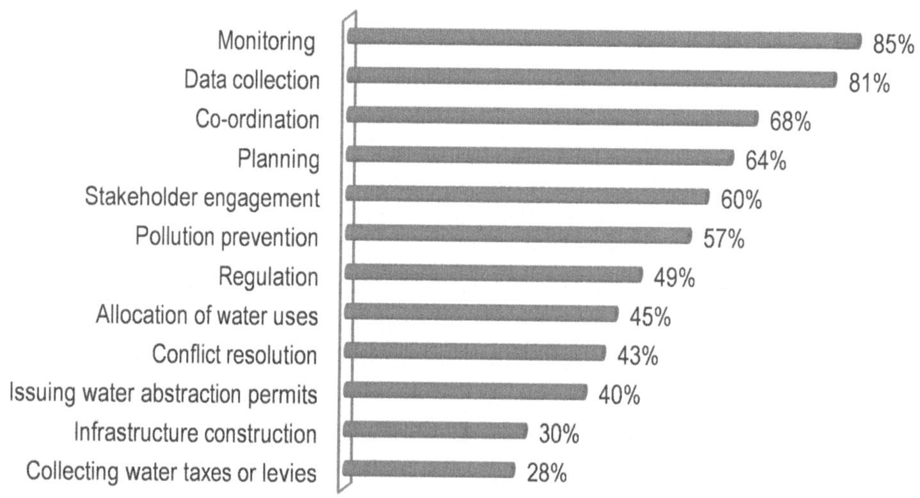

Note: Based on the 53 river basin organisations of surveyed cities.

Source: OECD Survey on Water Governance for Future Cities (2014).

Service providers

Service providers are key actors in the urban water management system. Amongst surveyed cities, the highest number of service providers for drinking water and sanitation services is reported in the Chinese cities with 15 operators in Suzhou and 22 in Zibo. In most cases there is only one (67%) service provider within the city's administrative perimeters (Figure 3.8).

When analysing by water function, a greater number of service providers is reported in the case of wastewater treatment (Figure 3.9). Drinking water supply, sewage collection and wastewater treatment services are bundled within the same service provider in 67% of surveyed cities. Large municipal systems can improve a water system's ability to finance needed investments while simultaneously increasing efficiency and cost-effectiveness. Conversely, the risk of diseconomies of scale may rise in megacities since high costs are attached to water transportation and network maintenance. The question of the appropriate operational scale depends on context, and there is not a unique solution (Box 3.2). In most cases, it may differ for drinking and non-potable water for example and require unbundled services, which best fit the purpose, while making the best use of scale and scope effects (OECD, 2009a).

Figure 3.8. **Number of service providers in surveyed cities**

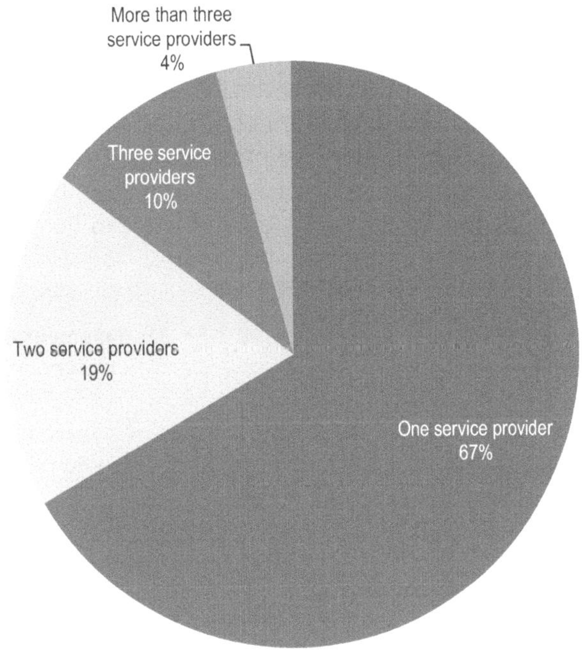

Note: Number of service providers operating in cities' administrative boundaries for drinking water supply, sewage collection and wastewater treatment, according to data provided by the 48 surveyed cities.

Source: OECD Survey on Water Governance for Future Cities (2014).

Figure 3.9. **Number of service providers by water functions within cities' administrative perimeters**

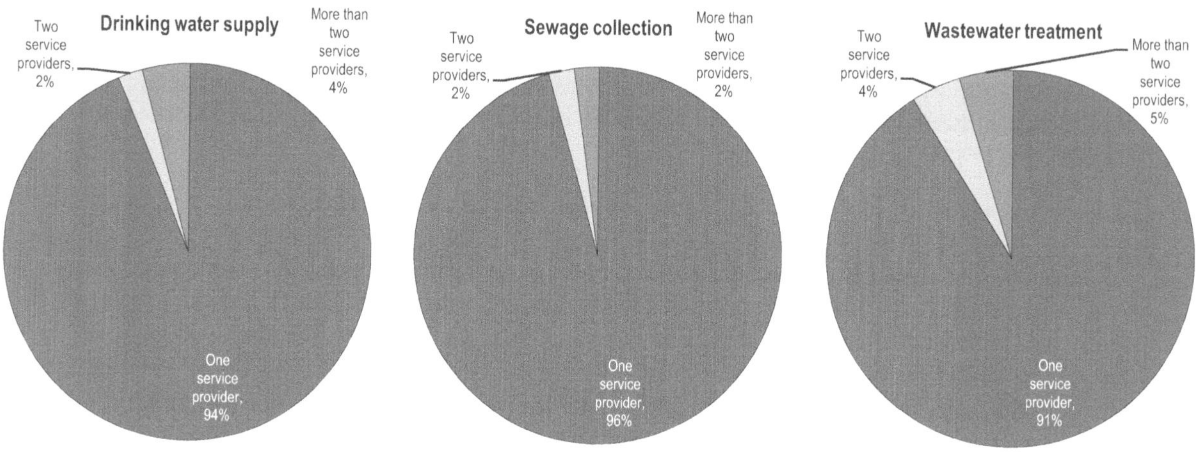

Note: Based on answers from the 47 cities surveyed for drinking water supply and 45 cities in the case of sewage collection and 46 in the case of wastewater treatment.

Source: OECD Survey on Water Governance for Future Cities (2014).

There is a variety of management models for drinking water and sanitation services across cities (Figures 3.10 and 3.11). According to the literature (Marques, 2010; OECD, 2015a), either the public sector is responsible for the management of water services and owns the assets, or assets and management are both in private hands (United Kingdom). In the first case, there are several options for provision: i) it can be delegated to public water operators (Austria, Germany, Netherlands); ii) it can be delegated to private companies following public tenders (France); or iii) there can be cases of public-private partnerships (PPPs) (Italy) (Figure 3.12).

Figure 3.10. **Management models for bundled water services**

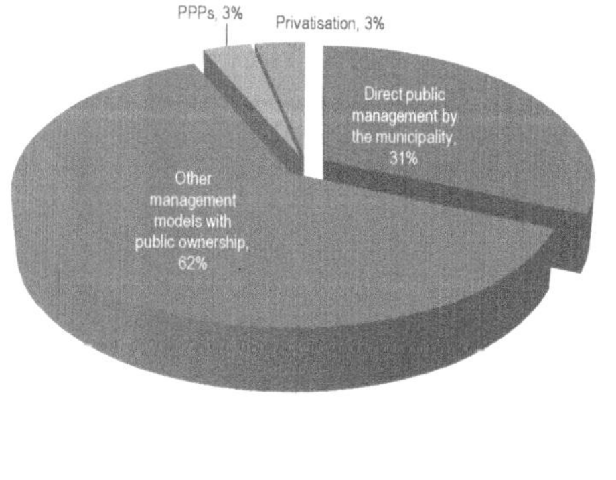

Note: Based on the answers of 29 cities where water services are bundled and that indicated the management model.

Source: OECD Survey on Water Governance for Future Cities (2014).

Figure 3.11. **Management models for unbundled water services**

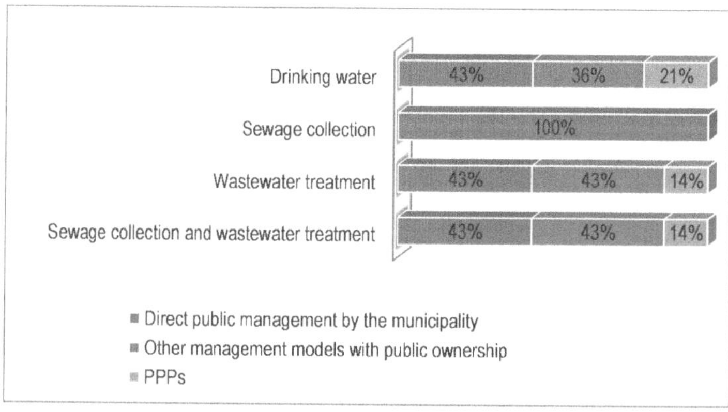

Note: Based on the answers provided by respondents. Percentages correspond to 14 cases for drinking water; 6 cases for sewage collection, 7 for wastewater treatment and 7 cases for sewage collection and wastewater treatment together.

Source: OECD Survey on Water Governance for Future Cities (2014).

Box 3.2. **Searching for the appropriate operational scale for water services: A literature review**

Most OECD countries are concerned with the question of the "relevant scale" for public services, but an optimal size for water services cannot be identified and the literature shows mixed results:

- In the Netherlands, 500 000 households seem an optimum level for drinking water companies (House of Parliament, 2013 in OECD, 2014). De Witte and Dijkgraaf (2010) studied the effects of mergers in the Dutch drinking water sector, concluding the absence of significant impacts of mergers on efficiency within the water companies. The Administrative Agreement on Water Affairs signed in 2011 between the Ministry of Infrastructure and Environment, regional water authorities, drinking water companies, provinces and municipalities, aimed to foster efficiency gains across the water chain through improved collaboration (i.e. up to EUR 750 million per year by 2020); for example, better co-ordination between urban drainage and urban planning, more cost-effective organisation of wastewater treatment, ranging from further co-operation between municipalities and regional water authorities in the collection, transport and treatment of wastewater, to joint collection of taxes, etc.
- A study covering 55 French water utilities for the years 1995 to 1997 shows that profitability is highest when a water district is made of up to five communities (Garcia and Thomas, 2001).
- In Italy, the definitions of optimal territorial areas (ATOs) helped achieve economies of scale (OECD 2013a). The ATO governing body, defined by the Region, is in charge of choosing the organizational form underlying water operations, elaborating the investment programs and defining tariffs, amongst others. The consolidation of water service operators took place after the reform in 2006. Nowadays there are about 2 000 water service operators (OECD, 2015a).
- In Portugal, in 2015, reform for ensuring future financial sustainability pushed towards the mergers of 15 state-owned regional bulk water and wastewater utilities into three regional companies – North, Central litoral, Tagus Valley-Alentejo. De Witte and Marques (2011) found that Portugal optimally would count about 60 utilities (in contrast to the 300 utilities operating in 2008).
- In Germany the optimal size of a water supply utility would be about 66 000 inhabitants (18 500 connections), according to a study carried out by Graetz (2008). Zschille (2014a, 2014b) highlights the difficulty of drawing conclusions on the benefits of mergers in Germany. Typically studies do not take into account detailed consideration of the companies with regard to their external influences, such as the characteristics of coverage areas, or considerations of quality of care. Furthermore, it would be necessary to include potential environmental aspects.
- In Japan, Urakami and Parker (2011), looking at the economic effects of consolidation of Japanese water utilities in the period 1999–2006, found that consolidation has had some beneficial impact on cost effectiveness, but that the result is limited.

Source: OECD (2014), Water Governance in the Netherlands: Fit for the Future? OECD Studies on Water, OECD Publishing, Paris, http://dx.doi.org/10.1787/9789264102637-en; OECD (2015a), The Governance of Water Regulators, OECD Studies on Water, OECD Publishing, Paris, http://dx.doi.org/10.1787/9789264231092-en.; OECD (2013a), OECD Environmental Performance Reviews: Italy 2013, OECD Publishing, Paris.http://dx.doi.org/10.1787/9789264186378-en; De Witte, K. and R. Marques (2011), "Big and beautiful? On non-parametrically measuring scale economies in non-convex technologies", Journal of Productivity Analysis, Volume 35/3), pp. 213–226; De Witte, K. and E. Dijkgraaf (2010), "Mean and bold? On separating merger economies from structural efficiency gains in the drinking water sector", Journal of the Operational Research Society, Volume 61/2, pp. 222–234; Garcia, S. and A. Thomas (2001), "The structure of municipal water supply costs: Application to a panel of French local communities", Journal of Productivity Analysis, July 2001, Volume 16/1, pp. 5-29; Graetz, H. (2008), "Synergiepotenzial einer fragmentierten Wasserwirtschaft Ein Beitrag zum Wert des Zusammenwirkens in fragmentierten Organisationsstrukturen der Wasserwirtschaft, Universiät Weimar", in Kommunalkredit Public Consulting (2009), "Report on measures to cope with over-fragmentation in the water supply and sanitation sector", final report prepared for the OECD, September, available at: http://www.publicconsulting.at/uploads/final_report_overfragmentation.pdf; Urakami, T. and D. Parker (2011), "The effects of consolidation amongst Japanese water utilities: A hedonic cost function analysis", Urban Studies, Volume 48/13, pp. 2805–2825; Zschille, M. (2014a), "Nonparametric measures of returns to scale: An application to German water supply", Empirical Economics, Volume 47/3, pp. 1029-1053; Zschille, M. (2014b), "Marktstrukturen in der Trinkwasserversorgung", DIW Berlin, http://www.diw.de/de/diw_01.c.487385.de/presse/diw_roundup/marktstrukturen_in_der_trinkwasserversorgung.html, (accessed December 2015).

Figure 3.12. **Management models for water supply and sanitation from a sample of surveyed cities**

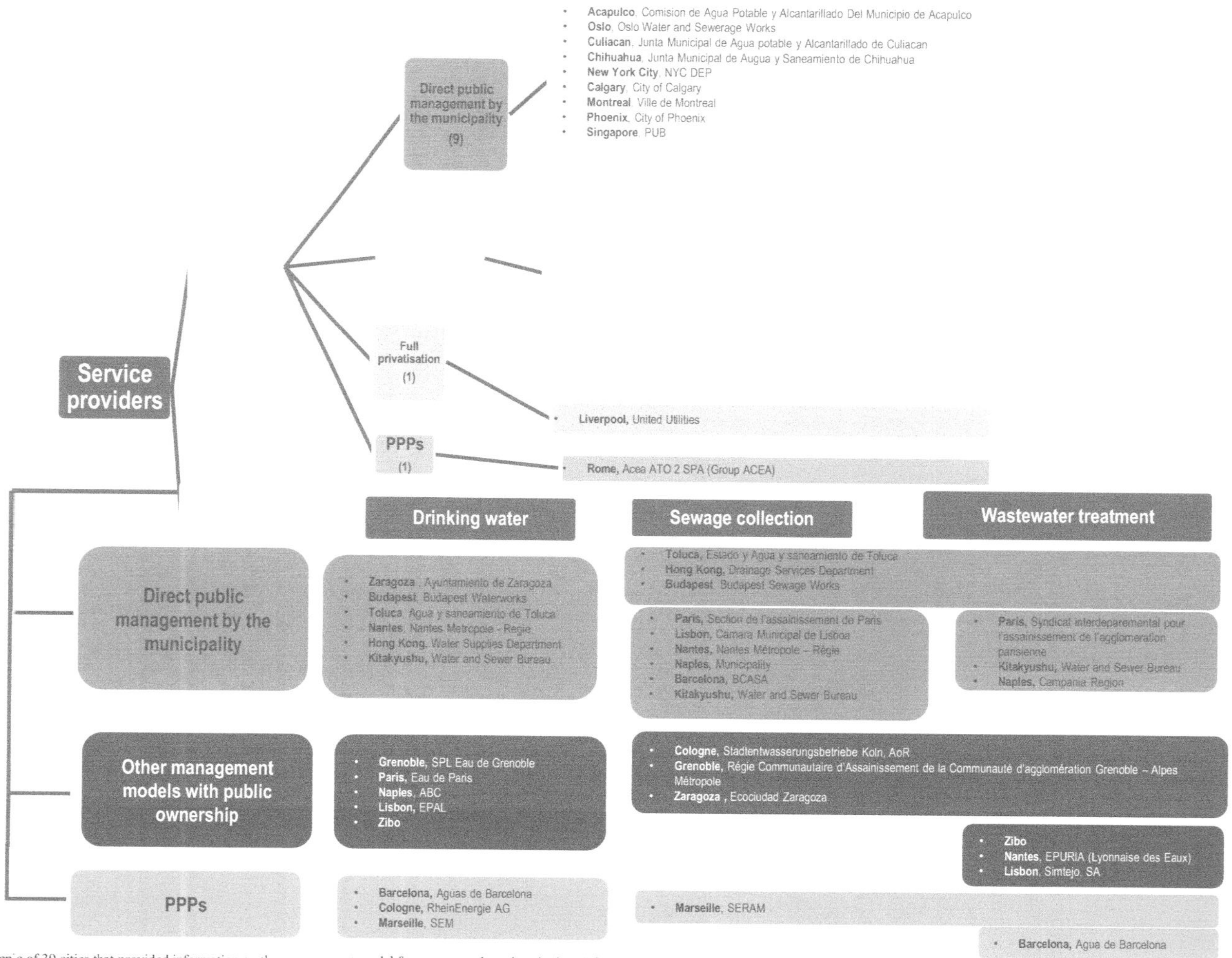

Note: Based on a samp e of 39 cities that provided information on the management model for water supply and sanitation. Information is missing for the cities of Rio de Janeiro, Daegu, Suzhou, Okayama and Queretaro.
Source: OECD Survey on Water Governance for Future Cities (2014).

In the case of bundled water services, in the majority of situations, public utilities are not directly managed by the municipalities (62%). They can be regulated under private law and have financial autonomy. The second case involves situations where water services are subject to direct public management by the municipalities (31%). The prevailing management model in the case of unbundled services is direct public management by the municipality for drinking water supply (43%) and sewage collection (100%), while there is not a prevailing form of management for wastewater treatment between direct public management by the municipality and other forms of public management (43% in both cases). Public-private-partnership contracts are less common and mostly used for drinking water supply[3] in cities surveyed (Table 3.1).

Table 3.1. **Examples of management models for water utilities**

Country	Management models
Czech Republic	In the Czech Republic, utilities work through municipality-owned assets and municipalities are the main responsible bodies.
Denmark	Municipalities hold 165 systems, while the number is wider for co-operatives (2 575).
France	There are 15 000 water utilities and 17 000 wastewater utilities. The share of private sector participation in water supply and sanitation management is high compared to OECD countries. The City of Paris re-municipalised in 2010 and other cities followed this example.
Germany	There is a great variety of water and wastewater management models in Germany; The municipal utilities (*Stadtwerke*) and publicly organised utilities (*Zweckverband* or *Anstalt öffentlichen Rechts*) have a legal autonomy. The rules differ from the federal states. Especially for the wastewater sector, in Germany there are special lawed water boards in North Rhine-Westphalia with autonomy. For the water sector there are also water and soil associations with a special law (*Gesetz über Wasser- und Bodenverbaende*) and autonomy.
Greece	The two mixed entities (EYDAP and EYATH) supply the majority of the population followed by municipal companies and municipalities through direct management models. EYDAP is a member of the Athens Stock Exchange (IPO in 2000), with the following shareholders: Greek State (34%), Hellenic Republic Asset Development Fund (HRADF) (27%), Paulson and Co (10%), other (29%). EYDAP is managed by the Board of Directors (BoD) which is elected in the General Shareholder Meeting. The majority shareholder, the Greek State, designates 9 of the 13 members of the BoD.
Italy	Some of the biggest Italian cities are managed by public operators (e.g. Milan, Turin, Venice, Naples). Examples of PPPs are in Rome and Florence.
Japan	In the majority of cases, the service provision is delivered by local public corporations. Municipal enterprises may have financial autonomy, as in the case of Tokyo Waterworks.
Korea	K-Water is a state-owned operator that manages half the national water supply facility capacity; operates 22 local water supply systems and participates in the sewerage business.
Mexico	Municipalities are responsible for providing water and sanitation, either directly (e.g. service providers that are part of the municipal government) or indirectly (e.g. providers that are legally separate entities wholly owned by the municipality). They can also delegate responsibility to private operators or utilities owned and operated by the state government. Water utilities, public, decentralised organisms (*Organismos Operadores*) are in charge of service delivery.
Netherlands	Drinking water companies (10) provide drinking water supply, operating under private law with public shareholders. Their geographic scope covers on average the area of two or three regional water authorities and between 20 and 50 municipalities. Regional water authorities (24) manage regional water systems to maintain water levels, water quality and wastewater treatment; they are decentralised public authorities endowed with specific legal personality and financial resources.
Norway	About 95% of the population is served by public drinking water supplies from over 2 000 waterworks. These include municipal, inter-municipal, state-owned and private waterworks.
Portugal	The water reform that started at 1993 led to the creation of 18 state-owned regional companies to produce quality bulk water and wastewater for municipalities. In 2015, there was a reorganisation that aggregated 15 state-owned regional bulk water and wastewater utilities into three regional companies – North, Central litoral, Tagus Valley- Alentejo – to bring economies of scale, allow tariffs to converge in a narrower band across the country.
Spain	Various actors provide water services under the responsibility of municipal governments.
United Kingdom	In England and Wales, OFWAT is the economic regulator. Its purpose is to guarantee quality service at a fair price from the ten regional companies in charge of water and wastewater management. Scottish Water is a publicly owned company, answerable to the Scottish Parliament and operating under specific regulatory requirements in relation to the operation of the retail market for commercial customers. Northern Ireland Water is a government-owned company set up in April 2007. The Utility Regulator is responsible for regulating the electricity, gas, water and sewerage industries in Northern Ireland.
United States	Most water service providers are municipal corporations. Tariff regulation is overseen by state public utility commissions. Penetration into the US market by private water companies has so far been limited. Most of the private operations are foreign owned. Investment is mostly by municipal bonds. Much of the wastewater infrastructure construction receives federal subsidy.

Table 3.1. **Examples of management models for water utilities** *(continued)*

Brazil	Municipalities are primarily responsible for service provision. In the case of the state of Rio de Janeiro, the state plays an important role in large urban agglomerations to exploit economies of scale and cross-subsidise poor municipalities. Service providers need to report back to the state regulatory agency on indicators and investments.
China	In China, water is managed by the public sector. Urban water supply and sanitation may be organised according to geographic and local administrative boundaries.

Source: Compilation by the authors from recent OECD publications, in particular: OECD (forthcoming), *OECD Territorial Reviews: Japan*, OECD Publishing, Paris; OECD (2015b), *Water Resources Governance in Brazil*, OECD Publishing, Paris, http://dx.doi.org/10.1787/9789264238121-en; OECD (2015c), *Water and Cities: Ensuring Sustainable Futures*, OECD Studies on Water, OECD Publishing, Paris, http://dx.doi.org/10.1787/9789264230149-en; OECD (2015d), *OECD Environmental Performance Reviews: Spain 2015*, OECD Publishing, Paris, http://dx.doi.org/10.1787/9789264226883-en; OECD (2014), *Water Governance in the Netherlands: Fit for the Future?*, OECD Studies on Water, OECD Publishing, Paris, http://dx.doi.org/10.1787/9789264102637-en; OECD (2013b), "Highlights from the 1st OECD Water Governance Initiative", 27-28 March 2013, available at: www.oecd.org/gov/regional-policy/OECD-WGI-1st-Meeting-highlights.pdf; OECD (2013c), *Making Water Reform Happen in Mexico*, OECD Studies on Water, OECD Publishing, Paris, http://dx.doi.org/10.1787/9789264187894-en; OECD (2011b), *OECD Environmental Performance Reviews: Norway 2011*, OECD Publishing, Paris, http://dx.doi.org/10.1787/9789264098473-en; OECD (2009b), *Reviews of Regulatory Reform: China: Defining the Boundary between the Market and the State*, OECD Publishing, OECD, Paris, http://dx.doi.org/10.1787/9789264059429-en; Kauffmann, C. (2012), "Framework conditions for private sector participation in water infrastructure in Mexico", available at: www.oecd.org/daf/inv/investmentfordevelopment/Checklist%20assessment%20of%20Mexico.pdf; Suez Environment and UGCL (2009), "Role of local government in water supply and sanitation", A global overview of governance models, www.suez-environnement.fr/brochure-uglc, (accessed in October 2015); Scottish Water website, Governance, www.scottishwater.co.uk, (accessed in October 2015); Utillity Regulator website, Who we are, www.uregni.gov.uk, (accessed in October 2015); Northern Ireland Water website, www.niwater.com, (accessed in October 2015); K-Water (n.d.), About K-Water, http://english.kwater.or.kr, (accessed in October 2015).

Conclusion

The institutional mapping exercise carried out in this chapter shed some light on the allocation of roles and responsibilities across levels of government concerning urban water management. The results show that sub-national authorities, in particular municipalities, play a key role in urban water management (drinking water supply, wastewater services, sewage collection, drainage to water security). In practice, however, responsibility is shared across levels of governments with a predominant role held in some cases by the central government. This division of responsibilities is due to the water challenges cities confront and associated need for co-ordination.

Full decentralisation of water governance at the city level may not be optimal or desirable. Basin-level management, for example, may require upper-level governance to avoid inequities in water allocation within a water basin and also to ensure that the public good aspects and values of water are given sufficient recognition (OECD, 2011a). At the same time, while inter-municipal, supra-municipal and metropolitan bodies have been set up to respond to co-ordination challenges while creating economies of scale and improving the quality of service, there might be risks of duplication and overlap of responsibilities with lower and higher levels of government.

Notes

1. Water security is about managing water risks, including risks of shortage, water excess, water pollution and risks of undermining the resilience of freshwater systems. The OECD developed a risk-based approach consisting of three steps: "knowing", "targeting" and "managing" (OECD, 2013d).

2. In New York City, for example, the water system operates according to rules and regulations established by the State of New York, which in turn, has promulgated its rules and regulations to be at least as stringent as required by the central authority (the US Government). The State and the Central authority have oversight, enforcement, and regulatory responsibilities but operational ones are held at local level.

3. A public-private partnership refers to a contractual agreement between the entity that has the overall legal responsibility for providing drinking water or wastewater services and a company with private status acting as operator with overall responsibility for the management of the service.

Bibliography

De Witte, K. and E. Dijkgraaf (2010), "Mean and bold? On separating merger economies from structural efficiency gains in the drinking water sector", *Journal of the Operational Research Society,* Volume 61/2, pp. 222–234.

De Witte, K. and R. Marques (2011), "Big and beautiful? On non-parametrically measuring scale economies in non-convex technologies", *Journal of Productivity Analysis,* Volume 35/3, pp. 213–226.

Garcia, S. and A. Thomas (2001), "The structure of municipal water supply costs: Application to a panel of French local communities", *Journal of Productivity Analysis*, July 2001, Volume 16/1, pp. 5-29.

Graetz, H. (2008), "Synergiepotenzial einer fragmentierten Wasserwirtschaft: Ein Beitrag zum Wert des Zusammenwirkens in fragmentierten Organisationsstrukturen der Wasserwirtschaft, Universität Weimar", in Kommunalkredit Public Consulting (2009), "Report on measures to cope with over-fragmentation in the water supply and sanitation sector", final report prepared for the OECD, September, available at: www.publicconsulting.at/uploads/final_report_overfragmentation.pdf.

Kauffmann, C. (2012), "Framework conditions for private sector participation in water infrastructure in Mexico", available at: www.oecd.org/daf/inv/investmentfordevelopment/Checklist%20assessment%20of%20Mexico.pdf.

K-Water (n.d.), "About K-Water", http://english.kwater.or.kr, (accessed in October 2015).

Marques (2010), *Regulation of Water and Wastewater Services – An International Comparison*, IWA Publishing, London.

Northern Ireland Water website (n.d), www.niwater.com, (accessed in October 2015).

OECD (forthcoming), *OECD Territorial Reviews: Japan*, OECD Publishing, Paris.

OECD (2015a), *The Governance of Water Regulators*, OECD Studies on Water, OECD Publishing, Paris, http://dx.doi.org/10.1787/9789264231092-en.

OECD (2015b), *Water Resources Governance in Brazil*, OECD Publishing, Paris, http://dx.doi.org/10.1787/9789264238121-en.

OECD (2015c), *Water and Cities: Ensuring Sustainable Futures*, OECD Studies on Water, OECD Publishing, Paris, http://dx.doi.org/10.1787/9789264230149-en.

OECD (2015d), *OECD Environmental Performance Reviews: Spain 2015*, OECD Publishing, Paris, http://dx.doi.org/10.1787/9789264226883-en.

OECD (2014), *Water Governance in the Netherlands: Fit for the Future?*, OECD Studies on Water, OECD Publishing, Paris, http://dx.doi.org/10.1787/9789264102637-en.

OECD (2013a), *OECD Environmental Performance Reviews: Italy 2013*, OECD Publishing, Paris, http://dx.doi.org/10.1787/9789264186378-en

OECD (2013b), "Highlights from the 1st OECD Water Governance Initiative", 27-28 March 2013, OECD, Paris, available at: www.oecd.org/gov/regional-policy/OECD-WGI-1st-Meeting-highlights.pdf.

OECD (2013c), *Making Water Reform Happen in Mexico*, OECD Studies on Water, OECD Publishing, Paris, http://dx.doi.org/10.1787/9789264187894-en.

OECD (2013d), *Water Security for Better Lives*, OECD Studies on Water, OECD Publishing, Paris, http://dx.doi.org/10.1787/9789264202405-en.

OECD (2011a), *Water Governance in OECD Countries: A Multi-level Approach*, OECD Studies on Water, OECD Publishing, Paris, http://dx.doi.org/10.1787/9789264119284-en.

OECD (2011b), *OECD Environmental Performance Reviews: Norway 2011*, OECD Publishing, Paris, http://dx.doi.org/10.1787/9789264098473-en.

OECD (2009a), *Managing Water for All: An OECD Perspective on Pricing and Financing*, OECD Studies on Water, OECD Publishing, Paris, http://dx.doi.org/10.1787/9789264059498-en.

OECD (2009b), *Reviews of Regulatory Reform: China: Defining the Boundary between the Market and the State*, OECD Publishing, Paris, http://dx.doi.org/10.1787/9789264059429-en.

Scottish Water website (n.d.), "Governance", www.scottishwater.co.uk, (accessed in October 2015).

Suez Environment and UGCL (2009), "Role of local government in water supply and sanitation", www.suez-environnement.fr/brochure-uglc, (accessed October 2015).

Urakami, T. and D. Parker (2011), "The effects of consolidation amongst Japanese water utilities: A hedonic cost function analysis", *Urban Studies* Volume 48/13, pp. 2 805-2 825.

Utillity Regulator website (n.d.), Who we are, www.uregni.gov.uk, (accessed in October 2015).

Zschille, M. (2014a), "Nonparametric measures of returns to scale: An application to German water supply", *Empirical Economics*, Volume 47/3, pp. 1 029-1 053.

Zschille, M. (2014b), "Marktstrukturen in der Trinkwasserversorgung", *DIW Berlin*, www.diw.de/de/diw_01.c.487385.de/presse/diw_roundup/marktstrukturen_in_der_trinkwasserversorgung.html, (accessed December 2015).

Chapter 4

Multi-level governance gaps in urban water management

This chapter identifies and analyses the primary governance bottlenecks cities face to effective water management within the context of the OECD Multi-level Governance Framework, "Mind the Gaps, Bridge the Gaps". Such gaps are related to questions of scale mismatch (administrative gap), silos and fragmentation (policy gap), diverging rationales and objectives (objective gap), asymmetries of information (information gap), lack of capacity (capacity gap), insufficient resources (funding gap), integrity and transparency (accountability gap). This chapter portrays their relative importance across the 48 cities surveyed and paves the way for policy responses suggested in Chapter 5.

Key messages

The multiplicity of actors involved in water policy design and implementation raises multi-level governance challenges when co-ordination mechanisms are not in place.

Multi-level governance gaps are more or less acute from one city to another. While the funding gap is a prominent issue across surveyed cities, large cities are more concerned with information dispersed across agencies; cities growing above the OECD average face challenges in terms of scale for investment; sprawl cities struggle more than compact ones to engage with stakeholders; and cities without metropolitan governance arrangements have higher concerns over fragmentation of water-related tasks.

Capacity is often the "Achilles' heel" of sub-national governments: many cities are facing technical and human resources gaps to efficiently manage water. The former relates to planning, quality information, monitoring and evaluation. The latter covers issues of staff, expertise and managerial capabilities.

The mismatch between hydrological and administrative boundaries calls for a functional approach to water resources management, which should consider the scale of integrated basin systems rather than municipal jurisdictions.

Greater institutional incentives for co-ordination and co-operation across local authorities and policy domains are needed to overcome policy discontinuity and short-term views.

Overall OECD countries have made much progress in terms of water quality and water quantity data, but they can do better for economic, financial and institutional water-related data.

Higher transparency and integrity would enhance trust in decision makers and water managers through greater stakeholder engagement and publicly accessible information on the performance of water governance systems and outcomes.

OECD's multi-level governance framework

Water management cuts across multiple scales, levels of government and policy areas. To manage water within a whole-of-government approach, cities need to pay particular attention to a number of governance gaps. The OECD Multi-level Governance Framework (OECD, 2011a) identifies seven "gaps" to effective water policy design and implementation (Figure 4.1). These gaps are intrinsically linked to, or exacerbated by, key features of the water sector (both local and global, capital intensive, fragmented, monopolistic, etc.). They relate to the mismatch between administrative and hydrological boundaries (administrative gap), silos and fragmentation (policy gap), diverging rationales and objectives (objective gap), asymmetries of information (information gap), lack of capacity (capacity gap), insufficient resources (funding gap), as well as integrity and transparency (accountability gap).

Diagnosing multi-level governance gaps is a primary step to overcoming obstacles and promoting more integrated water management. The evidence provided in this chapter stems from perception-based indicators, collected through the OECD Survey on Water Governance for Future Cities. Given the subjectivity of the responses, the aim is not to show the intensity of each gap, but rather whether a governance gap exists or not, and which gaps are more acute across cities of the sample.

Figure 4.1. **OECD Multi-level Governance Framework: Mind the Gaps, Bridge the Gaps**

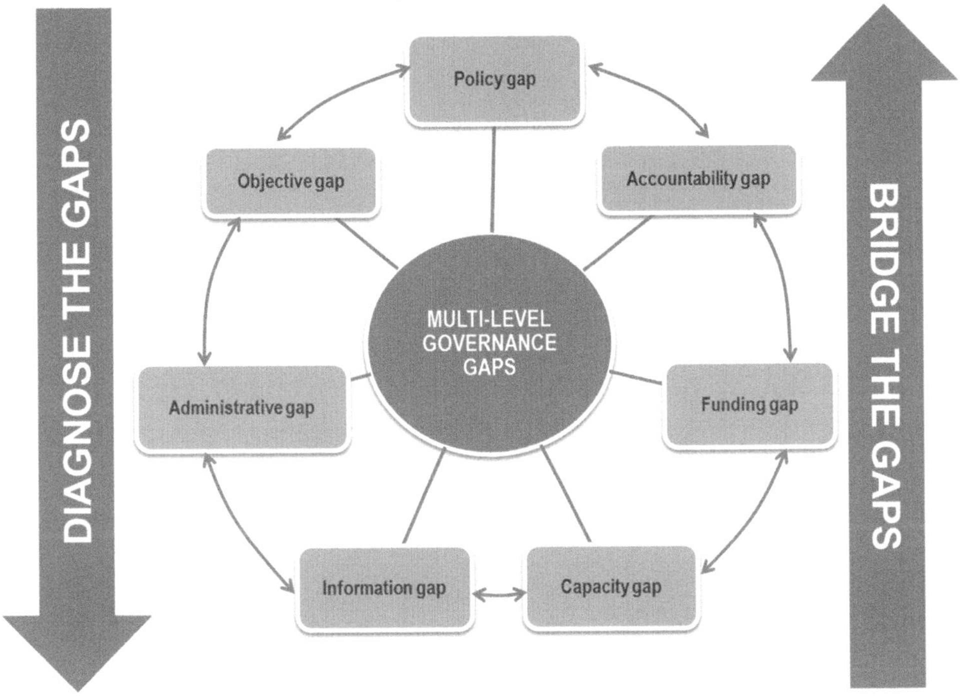

Source: OECD (2011a), *Water Governance in OECD Countries: A Multi-level Approach*, OECD Studies on Water, OECD Publishing, Paris, http://dx.doi.org/10.1787/9789264119284-en.

Scale

An administrative gap occurs when there is a mismatch between hydrological and administrative boundaries (Figure 4.2). Weak articulation between institutional, functional and hydrological logics affects urban water management because cities sit on watersheds, bounded hydrologic systems, which correspond to neither the administrative perimeter nor the functional areas. The blurred allocation of roles and responsibilities across multiple scales at sub-national level and their limited co-ordination can further exacerbate the administrative gap. Such a mismatch can hinder integrated urban management, for example when plans take into account limited jurisdictions (e.g. municipalities), overlooking a more effective territorial approach.

Cities are functional urban areas (FUAs) encompassing one or more small municipalities. They are defined by patterns of settlement and human activity (e.g. commuting) and often encompass multiple municipalities and their commuting zone (OECD, 2012a). In institutional terms, a *metropolitan city* corresponds to a specific form of governance arrangement for largely populated areas.

In the case of water, beyond the functional and institutional/administrative perspectives, the watershed, which follows hydrological logics, must be considered (Figure 4.3). As functional geographies depend on the function in question, in the case of water resources management, appraising the metropolitan and hydrological logics is key to addressing linkages between urban areas (where most people live) and the surrounding environments (rural and watersheds) that sustain them. This would optimise the opportunity cost of investments and the efficient use of water.

Figure 4.2. **Perceived scale mismatch for managing water in cities**

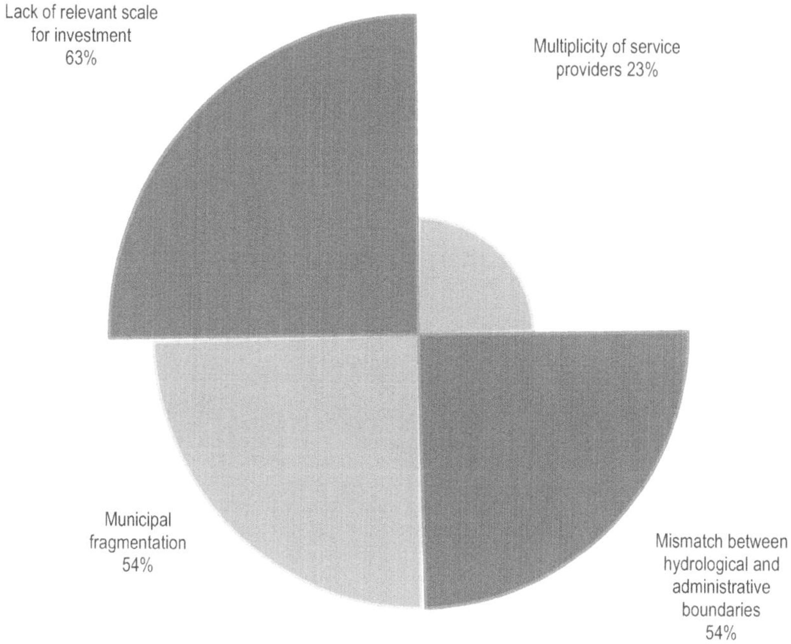

Note: Results based on a sample of 48 respondents who ranked the options provided as "major", "important" or "somewhat important" obstacle.

Source: OECD Survey on Water Governance for Future Cities (2014).

Figure 4.3. **Scales of water governance in cities**

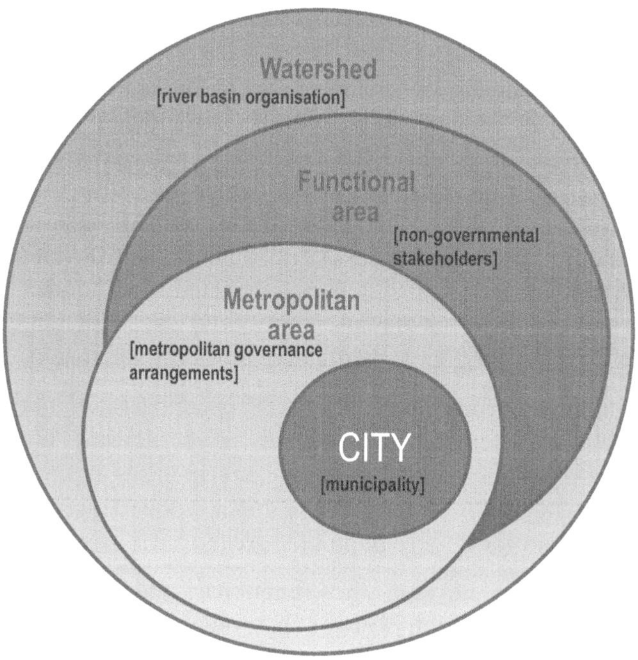

Source: Author's elaboration.

More particularly, a majority of cities surveyed identified the lack of relevant scale for investment as the most critical obstacle to effective urban water governance (63%) (Figure 4.2). Local or metropolitan governance seems to be appropriate for water supply and sanitation, while basin-level governance seems a better fit to water resources management, pollution, flood, drought and hydropower. River-basin organisations can intervene effectively in the physical system through an upstream-downstream "objective solidarity" among users for better stakeholder engagement in water policies (OECD, 2015a).

Municipal fragmentation, which occurs when there are multiple authorities at the local level, is a challenge in 54% of surveyed cities. Typically, this raises concerns about efficiency in the provision of service and the use of financial resources. Municipal fragmentation is very high in France: on average there are 57 municipalities per 100 000 inhabitants. This is one of the highest in the OECD (where the average is 10.6), together with the Czech Republic (59.4) and the Slovak Republic (54) (OECD, 2015b). Compared to other OECD countries, in 2012, France accounted for nearly 41% of all EU municipalities and 27% of those in the OECD area (OECD, 2014).

The administrative gap can have consequences in terms of competition over water uses and effectiveness of service delivery. In order to address water governance properly, there is a need to consider the territorial continuity of urban and peri-urban areas, according to a water-use based approach. Addressing the issue of scale may require revisiting the traditional criteria used to allocate natural resources (e.g. river basins) in relation to the administrative boundaries as well as related management instruments and the scale at which they operate.

Fragmentation and policy silos

A policy gap occurs when ministries, public agencies, authorities, departments work in silos without co-ordination mechanisms, and roles and responsibilities are not clearly allocated across levels of government. This has a negative impact on policy coherence and consistency in the absence of commitment to co-operate at all levels. As water holds implications for a number of related domains (spatial planning, agriculture, energy, etc.), co-ordination is essential to ensure a whole-of-government approach through which water can become a factor for sustainable growth and contribute to the broader economic, social and environmental agenda.

Cities reported that fragmentation of tasks and the lack of strategic vision across water-related sectors (54%) as the most prominent policy gaps to urban water governance (Figure 4.4). Similarly, unbalanced powers between authorities representing different interests (e.g. rural and urban areas) may hinder co-ordination (50%). The lack of legislative co-ordination (46%) may also impair water quality, because harmful legislation prevails over legal dispositions meant to prevent or remedy the problem. Overlapping or unclear responsibilities (46%) generate grey areas resulting in high transaction costs and delays.

Figure 4.4. **Perceived causes of institutional and policy silos**

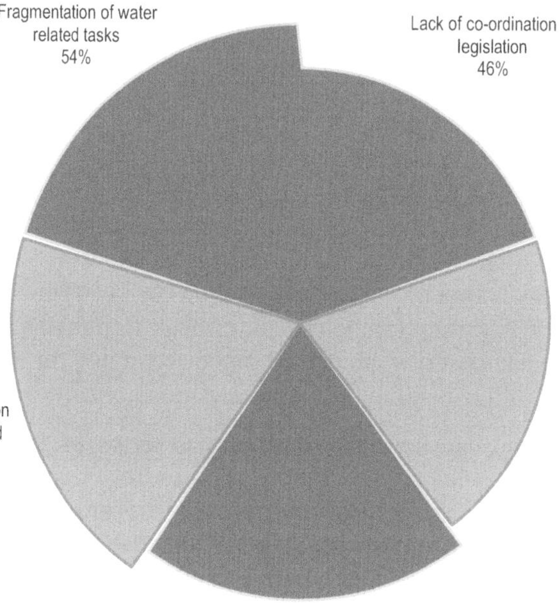

Note: Results based on a sample of 48 respondents who indicated the options provided represent a "major", "important" or "somewhat important" obstacle.

Source: OECD Survey on Water Governance for Future Cities (2014).

Diverging objectives compromising long-term management

An objective gap occurs when conflicting objectives and rationalities compromise long-term targets for integrated urban water policy and cross-sectoral co-ordination. This can happen either because of diverging interests between water-related fields or because of scarce vertical and horizontal co-ordination across levels of government. Cities surveyed reported the lack of institutional incentives for co-operation (65%) and contradictions between recommendations/directives at different levels of government (56%) as obstacles to long term and co-ordinated water policy to urban water governance. For instance, diverging objectives can take place between energy and water, as the generation of energy in some cases requires enormous consumption of water. Diverging objectives are likely to rise and cross-sectoral co-ordination is hindered when local authorities compete rather than collaborate (52%); when lobbies interfere bringing up sensitive interests (48%); and when conflicts over water allocation (46%) make it difficult to strike a balance between economic, social and environmental objectives (Figure 4.5).

Figure 4.5. **Perceived obstacles to long-term and co-ordinated water policy**

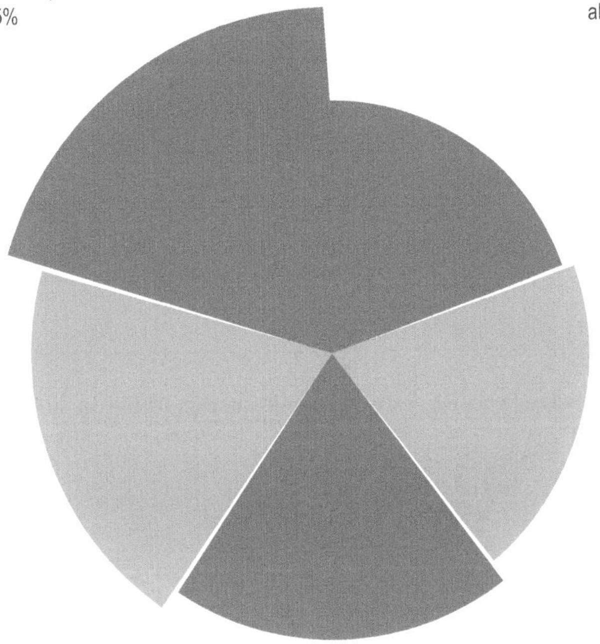

Note: Results based on a sample of 48 respondents who indicated the options provided represent a "major obstacle", "important obstacle" or "somewhat important" obstacle.

Source: OECD Survey on Water Governance for Future Cities (2014).

Financing

The funding gap consists in unstable or insufficient revenues undermining effective implementation of water responsibilities. Local governments' budgets have been further tightened by the financial crisis, with an increase of competing demands for limited public funds. Underinvestment can jeopardise the efficiency of urban water infrastructure in cities, thus compromising the capacity of institutions to deliver expected outcomes.

The ultimate sources of revenues in the water sector can be grouped into "3Ts", namely *tariffs* (revenues from the water bill), *taxes* (allocations from the public budget) and *transfers* from the international community (which have become secondary in most OECD countries) (OECD, 2011b). Revenues from these sources must increase in order to cover the costs of achieving agreed policy objectives now and in the future. Tariff structures and management models may need adjusting to secure stable revenues in the face of declining water consumption patterns. OECD (2015a) argues that a revision of the 3Ts model is necessary and other sources of funding need to be taken into account:

governments should consider levying taxes on those who benefit from increased water security (including land and property developers) or who generate higher costs and externalities (e.g. owners of large impervious surfaces, such as roads or car parks).

The funding gap has different origins, one of which being the difficulty in raising tariffs for water services in a context where costs are increasing (69%) (Figure 4.6). The financial crisis affected the ability for water utilities to raise tariffs in different ways: affordability constraints have hardened and the political power has been reluctant to increase tariffs (Box 4.1); borrowing has become more difficult. Financial management at the local level is often weak and local government entities' creditworthiness tends to be low (OECD, 2010). A total of 38% of cities surveyed do not have financial guarantees to borrow money. In others the "golden rule" of public finance restricts cities' capacity to borrow money from commercial sources of funding in order to finance investment. Other relevant funding issues, among the options provided in the survey, concern weak prioritisation of investment (48%) and difficulties in collecting tariffs and charges from water (46%).

Figure 4.6. **Perceived challenges to financial sustainability in surveyed cities**

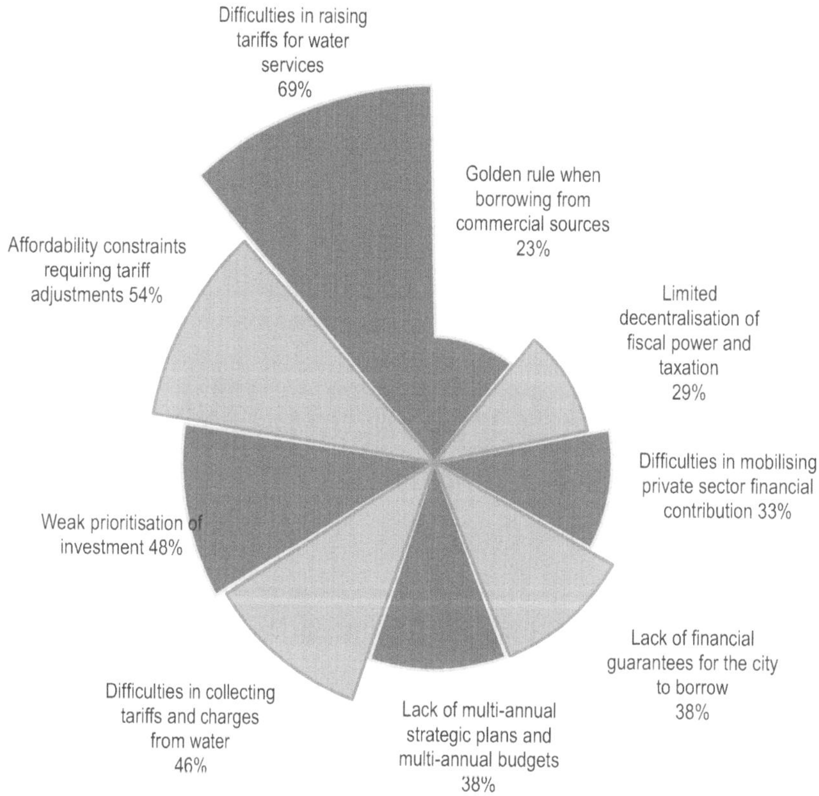

Note: Results based on a sample of 48 respondents who indicated the options provided represent a "major", "important" or "somewhat important" obstacle.

Source: OECD Survey on Water Governance for Future Cities (2014).

Box 4.1. **The challenge of charging for water in Budapest**

One of the most important challenges that the water sector, regulators and cities face in Budapest is the introduction of water utility tariffs, i.e. the price of drinking water, sewage collection, drainage and wastewater treatment.

According to present regulations, the public water utility supplier can apply prices set in 2013. A higher price can be introduced only in special cases foreseen by the Act and with the consent of the Authority. Therefore, the occurred surplus costs – whether it is allowable or not – cannot be built into the prices. The Minister for Water Utility is responsible for establishing the government-regulated prices and to determine the starting date of the application of these prices. The Hungarian regulator is only obliged to submit a proposal on tariffs each year to the minister, but delays can be observed in terms of approval. The Minister passes a Directive that takes into consideration the proposition of the Authority and determines the government-regulated price. As a result of the central price regulation, the sphere and rate of the allowable and occurred costs can be harmonised. The regulatory authority has the responsibility of supervising the price. Pricing can then be more transparent and pre-calculated for a medium- and long-term period, which should likely result in a reduction of the average price.

Source: Input from the City of Budapest for OECD Survey on Water Governance for Future Cities (2014).

Retail water tariff/user charges are among the economic instruments mostly in place for managing water in surveyed cities (88%). A majority also use fines, water pollution charges and levies. Payment for ecosystem services and marketable permits are less widely used (Figure 4.7).

Figure 4.7. **Economic instruments for urban water management in surveyed cities**

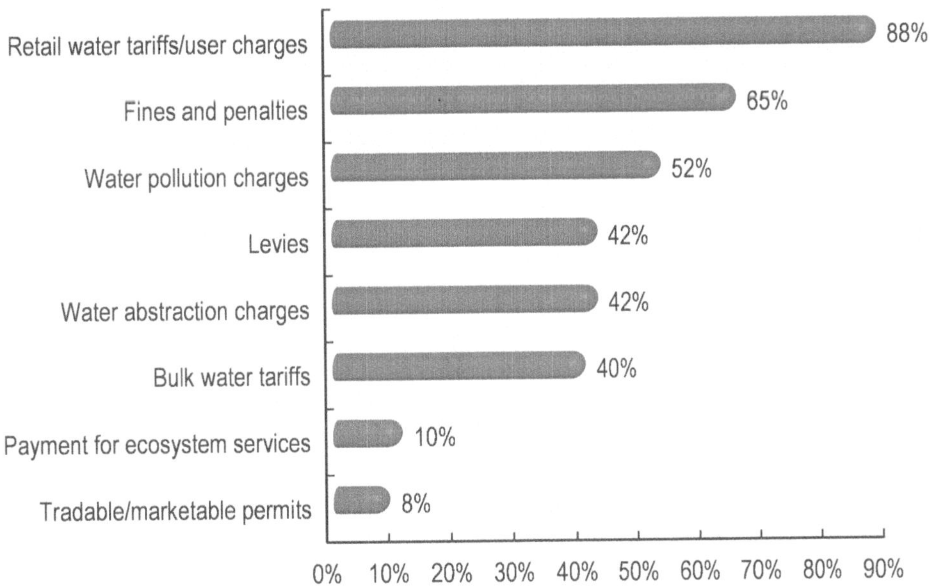

Note: Results based on a sample of 48 respondents who responded "yes" to the options provided.

Source: OECD Survey on Water Governance for Future Cities (2014).

A total of 64% surveyed cities reported that they play a role in tariff regulation (Table 4.1). The role of central vs. local governments in tariff setting reflects the degree of decentralisation of the sector in the country. In Mexico, the federal level has limited power to influence the process of setting tariffs at the local level as water services are a prerogative of municipalities by Constitution. The federal level can only provide guidance through the establishment of a voluntary norm and by providing financial incentives as part of federal programmes (OECD, 2013a). Tariff structures are set by each municipality (or by the state), according to the laws that apply to each federal state. In Germany, water tariffs are set at local level through different procedures in each *länder*. At federal level, the supervisory office for tariffs (Kommunalaufsichtsbehörde) has the responsibility to control the tariffs. In the United States, tariff regulation is overseen by state public utility commissions. In the Czech Republic, tariffs are set by individual utilities, but are subject to price controls by the Ministry of Finance.

Table 4.1. **Roles of local governments in water services' tariff regulation**

	Role at local level in tariff regulation
Acapulco	Approval of the tariffs
Amsterdam	Retail drinking water tariffs
Athens	Between 2000 and 2014 water and sanitation tariffs were defined jointly by the Ministry of Infrastructure, Transportation and Networks and the Ministry of Finance, based on proposals by the EYDAP's Board of Directors. Since 1 July 2015, such tariffs have been placed under the decisions of the Special Secretariat for Water in consultation with the Minister of Finance and other pertinent bodies. Decisions related the pricing policy for the different categories of users are now in effect for five years and issued at the end of each period for the following five-year period.
Bologna	City is represented in the ATERSIR that sets tariffs
Calgary	Setting up of the tariffs to be approved by City Council
Cologne	Decisions in terms of sanitation
Copenhagen	Approval based on national set price cap
Grenoble	The *Régie de l'eau potable* and *Société publique locale Eau de Grenoble* help calculate the price of water, aiding in the harmonisation of the price itself within the area
Hong Kong, China	Waterworks Ordinance and Sewage Services Ordinance
Lisbon	EPAL, involved with the drinking tariff and submits a proposal to the regulator
Mexico City	Setting up of tariffs to be approved
Milan	Setting up of tariffs by the *Ufficio d'ambito* –ATO of the City of Milan to be approved by the national regulator
Montreal	Setting up of tariffs, invoice and collection
Nantes	All tariffs for water and sanitation voted by the Metropolitan Community Council (water pricing, subscriptions water, water connection, works in sanitation)
New York	The rates system is adopted annually by the Water Board. It consists of seven members, each of whom is appointed by the New York City Mayor
Oslo	Full cost calculation for determining the tariffs
Paris	Setting royalties for the provision of drinking water and sanitation
Phoenix	Legislative authority to set tariffs
Rio de Janeiro	Role in sanitation (not specified)
San Luis Potosi	INTERAPAS proposes the amount of the tariffs for authorisation by the State Congress
Singapore	Tariffs/fines and penalties/levies
Stockholm	Set up of principles of tariff by the governing board of the City of Stockholm. Some basic principles of tariffs are: self-cost (affordable), sustainable financing of water management and encouraging local storm water treatment.
Queretaro	The State Water Commission (*Comision Estatal de Agua: CEA Querétaro*) proposes the level of tariffs to be approved by the *Junta de Consejo Consultivo de la Comisión Estatal de Aguas.*
Toluca	Setting up of tariffs to be approved
Zaragoza	Approval of the tariffs

Source: OECD Survey on Water Governance for Future Cities (2014).

In some countries, the basic principles for tariff structures and levels are defined at the national level including the number of tariff blocks, while leaving room for flexibility so that local, technical and social conditions can be taken into account. This is the case in Portugal where tariff setting is a shared competence of the national regulator and municipalities (OECD, 2015c). For instance, in Lisbon, EPAL is involved with the drinking tariff and submits a proposal to the regulator. Another example is the *Ufficio d'ambito* of Milan, which sets the tariff to be approved by the national regulator. In the United Kingdom, companies propose the tariffs. OFWAT reviews company plans on a five-year cycle and set maximum prices to encourage efficiencies and drive down costs.[1]

There are also different examples of tariff setting at metropolitan level: the HOFOR, a multi-supply company for metropolitan Copenhagen, which includes eight municipalities, is responsible for setting the tariff subject to a national set price cap; in Nantes, France all tariffs for water and sanitation are voted by the Metropolitan Community Council (i.e. water pricing, subscriptions water, water connection, works in sanitation); the Metropolitan Area of Barcelona in Spain sets water tariffs for the municipalities of the Barcelona conurbation. In many countries, policy makers and stakeholders face the difficulty of comparing prices due to the lack of harmonisation across municipalities in their definition and to the intrinsically localised nature of water resources and their distribution context.

The average price of water per household (p/m3) varies across cities. It ranges from 0.3 in San Luis Potosi to 6.8 in Kitakyushu (USD value in constant prices, constant PPP) (Figure 4.8). However, this variety is strongly related to a number of factors such as water consumption patterns, the distance to the withdrawal sources the status of the treatment plants and water networks, the investment required, the upgrading of collecting systems, the source of water (whether groundwater and then pumped or surface water, which may need treatment), the level of sprawling or compactness of a city.

Figure 4.8. **Average price of water per household in surveyed cities**

USD value in constant prices, constant PPP

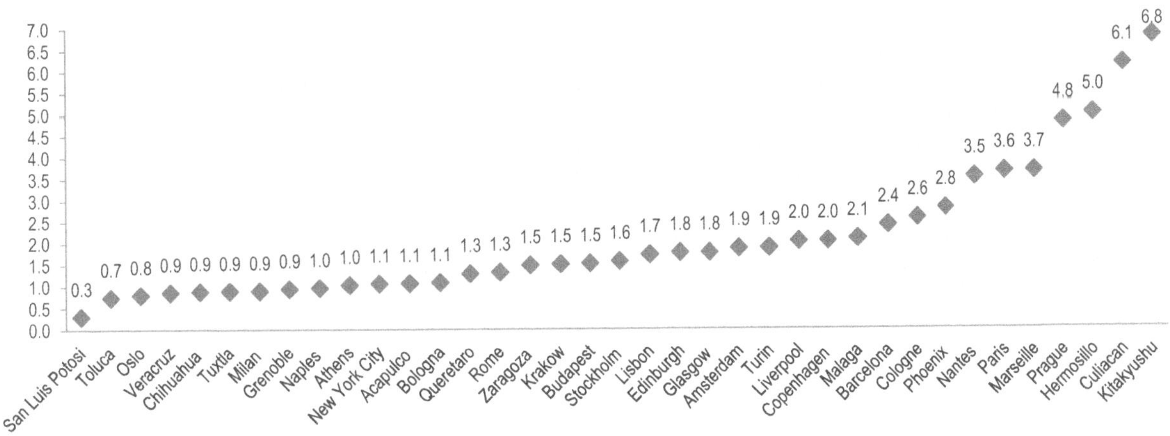

Note: USD value in constant prices, constant PPP (2010 reference year), based on the average price of water for households (price of a m^3 of domestic water) as reported by surveyed cities in local currency (year 2012). In the case of Stockholm the price is calculated on the average between the price of a household living in a detached (villa) house and that for a household in a flat. These prices are generated by the Swedish Water and Waste organisation. For Budapest and Queretaro prices are related to the year 2013.

Source: OECD Survey on Water Governance for Future Cities (2014).

The average price per cubic metre in itself therefore cannot allow for a performance comparison across cities. Still, it has been observed that prices are lower and the shortfall to full cost recovery usually greater in countries where water is scarce (OECD, 2013b). Water pricing has to take into account environmental sustainability, economic efficiency, financial sustainability, and social concerns, but price signals to users (households, industry and agriculture) should be transparent, clearer and along the principle of sustainable cost recovery (OECD, 2013b).

Many OECD cities have introduced measures to make water more affordable to the population at large and particularly to targeted groups (Figure 4.9). Solutions to tackle affordability, beyond modifying tariff structures, include providing income support (i.e. to compensate poor households for increases in the prices of services of public interest that are judged to be unacceptably burdensome) and facilitating payment (i.e. to help poor consumers manage their budgets by paying water bills at short intervals for example) (OECD, 2012b). In surveyed cities, such measures usually target poor populations (71%), disabled people (48%), populations living in disadvantaged areas (40%), and ethnic minorities (31%). They mostly consist in using progressive social tariffs (e.g. Grenoble, Hermosillo, Lisbon); avoiding water disconnection (e.g. Edinburgh, Glasgow); special tariffs for the disabled (e.g. Chihuahua, Nantes) and large families (e.g. Lisbon); assistance and pro-poor policies (e.g. Budapest, Calgary, Hong Kong, China) and for rural communities (e.g. Veracruz); grants for low-income families (e.g. Singapore);[2] or social funds for people living in disadvantaged areas (e.g. Grenoble and Malaga) or in relation to housing policy (Paris).

Figure 4.9. **Targeted groups benefitting from measures to guarantee access to water services**

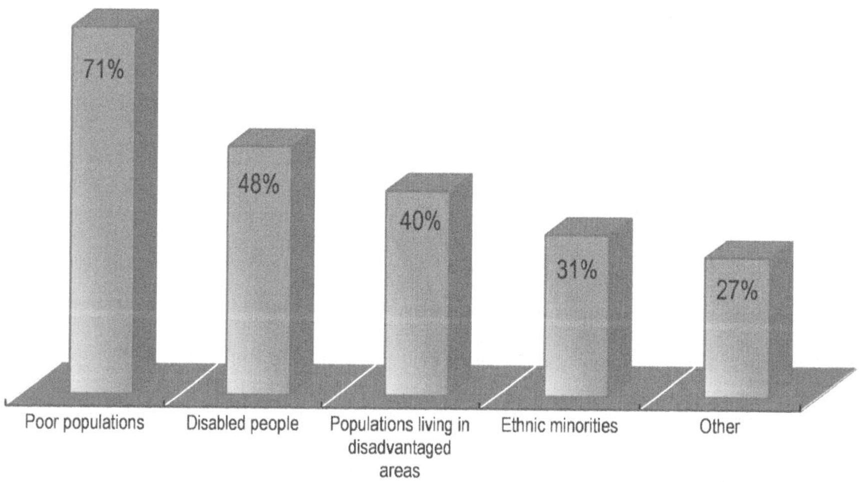

Note: Results based on a sample of 48 respondents who responded "yes" to the options provided.

Source: OECD Survey on Water Governance for Future Cities (2014).

Capacity and information

The capacity gap consists in the lack of scientific, technical, and infrastructural capacity of local actors, hindering the design and implementation of water policies. It often generates a vicious circle that triggers an information gap (quantity, quality, type), which in turn can generate an accountability gap (i.e. lack of transparency and integrity).

Human capacities and expertise are paramount to water governance at the city level. New governance arrangements and territorial reforms might be a challenge for sub-national level capacity. It is easier for established sub-national governments with well-developed institutions to take on new responsibilities than in cases where sub-national governments or related institutions must be created from scratch or historically have had a limited role.

Many cities in OECD countries are likely to face capacity gaps to manage water properly. This happens in particular when utilities are operating in the red; when more and more stringent environmental regulation cannot be enforced at lower level; when access to technological innovations is too costly; when the water sector does not attract sufficient professionals; and when systems are not in place to produce, use and share policy-relevant data for decision making and transparency purposes, etc.

Cities surveyed reported the lack of staff and managerial competencies (65%) as the main source of the capacity gap (Figure 4.10). Water in cities involves expertise from different fields and requires capacity to respond to emergencies (such as in cases of water-related extreme events), to set up measures for disaster prevention, as well as to carry out ordinary duties, coherently with citizens' needs, interconnection with other policies and sectors.

Figure 4.10. **Perceived capacity challenges to urban water management**

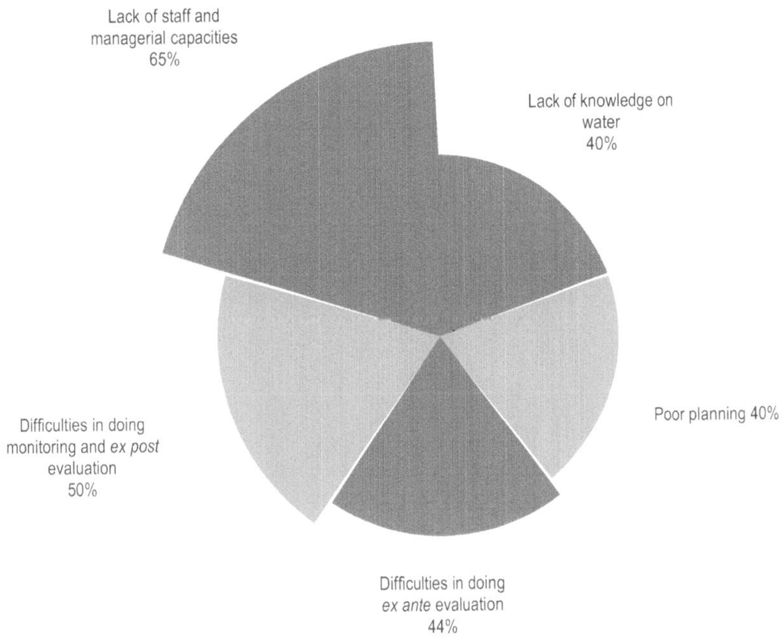

Note: Results based on a sample of 48 respondents who indicated the options provided represent a "major", "important" or "somewhat important" obstacle.

Source: OECD Survey on Water Governance for Future Cities (2014).

The information gap stems, amongst others, from the lack of or insufficiently robust water and water-related data and information to guide decisions across levels of government. Adequate information generation and sharing among relevant actors are of primary importance in order to overcome unclear definitions, incoherence across water functions and negligence due to lack of data. Data related to both operational management of the resource (e.g. hydro-geological, meteorological and hydrological data, service coverage data, records of consumption, cost recovery data, health impact data, operation and maintenance data, water quality data, etc.) and administrative infrastructure (e.g. including records relating to personnel, equipment, stores and finance). Overall OECD countries have made much progress in terms of water quality and water quantity data, as opposed to economic, financial and institutional dimensions of water management (OECD, 2011a).

The capacity to produce, collect and share data of good quality varies from one city to another. Depending on the purpose (e.g. setting water management plans, budgetary simulations, regulation, warning systems, etc.), data might be collected by local authorities, service providers, statistical offices or environmental agencies. Among surveyed cities, key issues concerning data usually concern the *what* (i.e. available information is too technical, for 56% of respondents), the *when* (i.e. data collection is incomplete and irregular, for 56% of respondents) and the *how* (i.e. data is dispersed across agencies making it difficult to track and compare; according to 52% of respondents) (Figure 4.11). The CEA Waters State Commission of Queretaro, the public institution providing drinking water and sanitation services, is organised in different operative areas (e.g. Sanitation Division), each of which has to generate a monthly statistical report at state and city level.

Figure 4.11. **Perceived information challenges to urban water management**

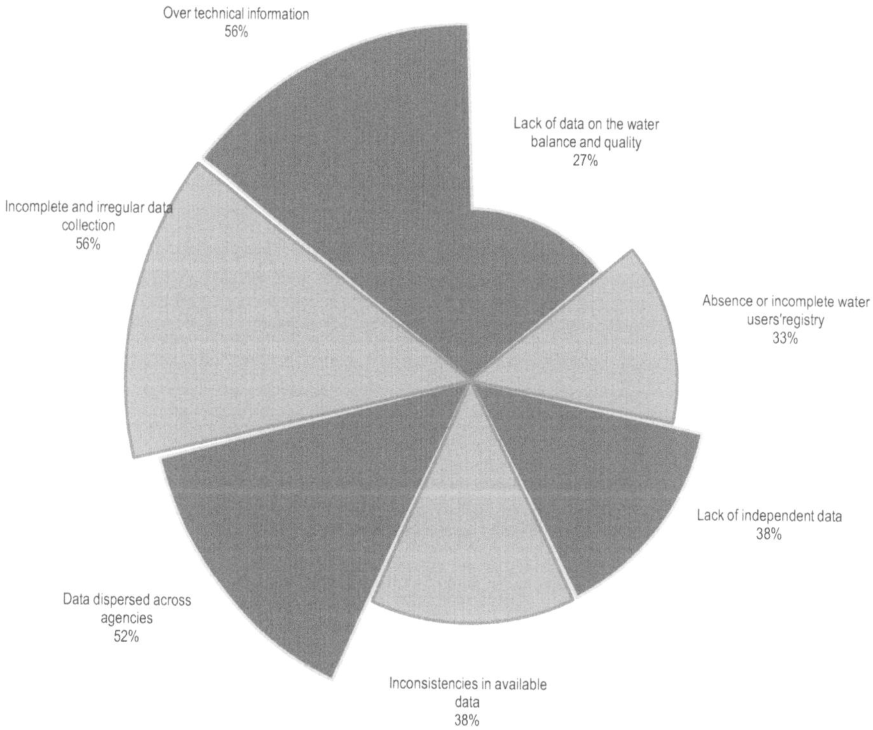

Note: Results based on a sample of 48 respondents who indicated the options provided represent a "major", "important" or "somewhat important" obstacle.

Source: OECD Survey on Water Governance for Future Cities (2014).

Accountability

An accountability gap arises when there is difficulty to ensure the transparency of practices across different constituencies. One of the causes can be asymmetric information across levels of government. Sub-national governments may have valuable information about local needs, preferences, policies and costs in the water sector, which might not be shared with other levels of government or be accessible to stakeholders. Asymmetries of information on the status of water supply and sanitation assets may lead to suboptimal contractual arrangements between the authority and the service provider, market abuse, and mistrust amongst consumers. Limited information sharing was reported as an obstacle by the majority of respondents (60%) (Figure 4.12). Easy and transparent access to data could address this challenge, which in addition to monopolistic behaviours, also justifies the recourse to regulatory instruments to protect the public interest (OECD, 2015c). Cities face many gaps hindering the transparency and participatory nature or urban water management, including weak stakeholder engagement (48%).

Figure 4.12. **Perceived transparency and accountability challenges to urban water management**

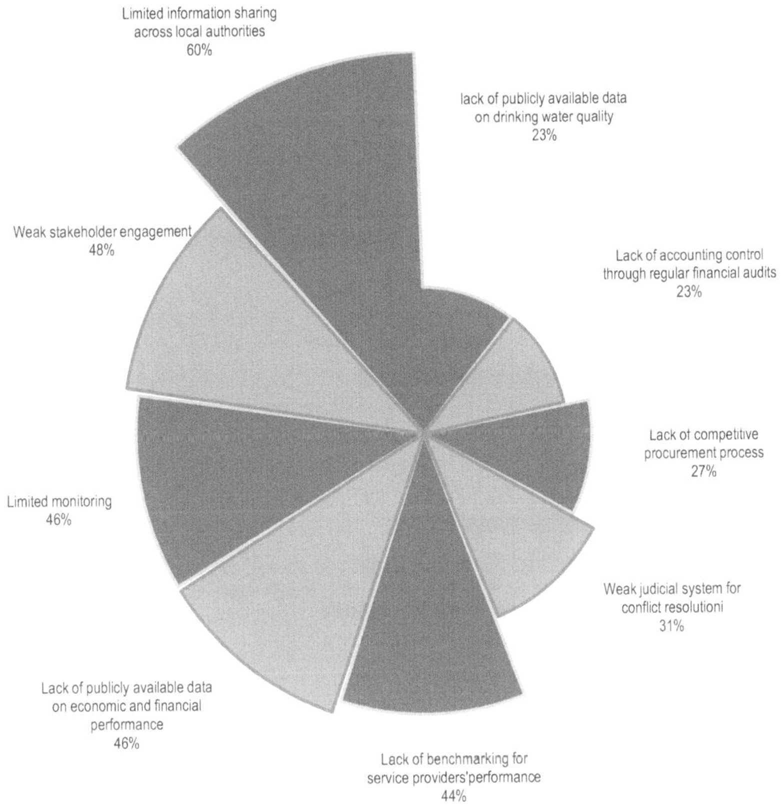

Note: Results based on a sample of 48 respondents who indicated the options provided represent a "major", "important" or "somewhat important" obstacle.

Source: OECD Survey on Water Governance for Future Cities (2014).

Limited monitoring and/or evaluation represent an obstacle for 46% of surveyed cities. In some cases, this may result from a lack of human and financial resources and expertise, and from a lack of data and available information. The OECD Survey on Water Governance for Future Cities shows that where they exist, monitoring and evaluation tools are frequently used (Figure 4.13). Benchmarks and evaluation reports allow competition-by-comparison and provide useful information to citizens, measuring the performance against indicators even if the former do not document opportunity costs. Financial analyses help to get a clear picture of the financial needs of the local government. National observatories can monitor service delivery performance and improve transparency of information and accountability of the water service sector. A key question is how well these instruments can be applied in guiding decisions.

Figure 4.13. **Existing monitoring and evaluation instruments**

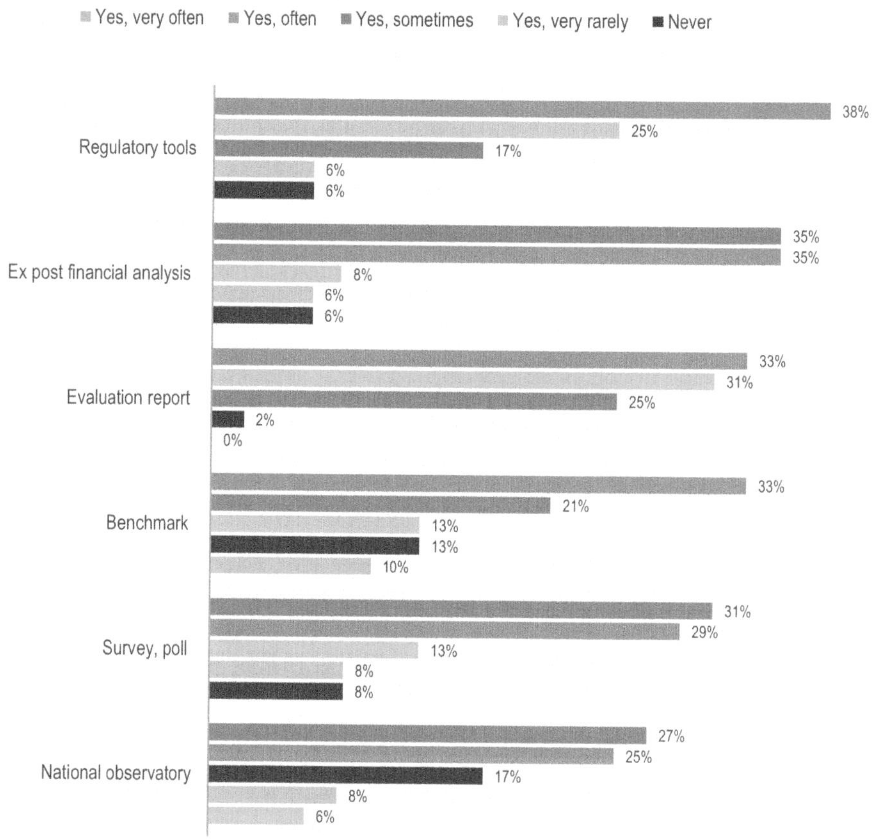

Note: Results based on a sample of 48 respondents who responded "yes" to the options provided.

Source: OECD Survey on Water Governance for Future Cities (2014).

A great variety of indicators exists for measuring the performance of water utilities, set up by third-party actors such as international organisations, regulators, or umbrella associations (OECD, 2015d).

- At international level, the International Benchmarking Network for Water and Sanitation Utilities (IBNET) database provides water utility performance measurements,[3] indicating utilities' health, stage of development, performance status, and the probability that a water utility will experience a performance problem (World Bank, 2014).

- The IWA (International Water Association) performance indicator project allows for performance assessment and benchmarking of water services, in terms of staff, equipment, operation, service quality and financial indicators (Cabrera, Jr. et al., 2011; OECD, 2015b).
- In the Netherlands, the Association of Water Companies (Vewin) publishes a yearly benchmarking report comparing the performance indicators of the public water companies.
- In Portugal, the regulatory agency ERSAR set up a compelling framework for measuring the performance of water utilities.
- In Austria, ÖWAV (Austrian Water and Waste management Association) and ÖVGW (Austrian Association for Gas and Water) are responsible of benchmarking.

Drawing on the Survey on Water Governance for Future Cities results, Figure 4.14 offers a snapshot of the number of employees per 1 000 connections, as reported by respondents. It is worth mentioning that such an indicator cannot, on its own, account for the complexity of the geographical, social and managerial contexts, nor be used as a proxy of efficiency or adequacy of human resources capacity at the local level (e.g. in this case the indicator should be weighted with other variables, such as the amount of turnover that is outsourced).

Figure 4.14. **Average number of water utilities' employees per 1 000 connections**

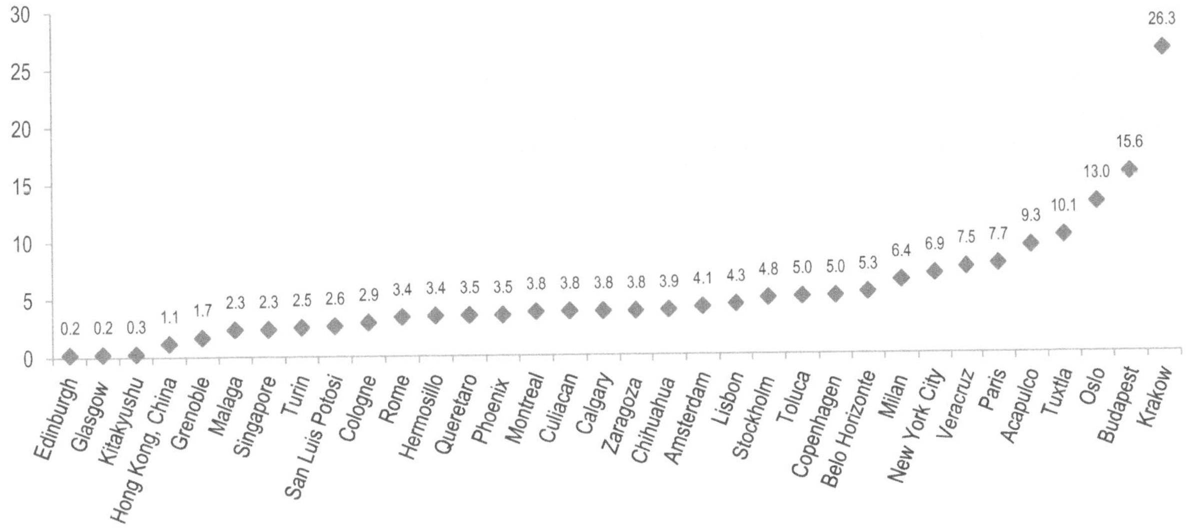

Note: In the case of unbundled services, the average number of employees and the average number of connections for each service (drinking water, sewage collection and wastewater treatment) have been calculated.

Source: OECD Survey on Water Governance for Future Cities (2014).

A summary of multi-level governance gaps by types of surveyed cities

The multi-level governance gaps analysed in this chapter are more or less acute across the types of cities, which may be more or less sensitive to them, as reported in Table 4.2.

Governance gaps differ when taking into account the **size** of cities.

- Surveyed cities with more than 5 million inhabitants are mostly concerned[4] with the lack of staff (capacity gap), limited information sharing (accountability gap), and data dispersed across agencies (information gap).
- Surveyed cities with a population between 1.5 million and 5 million inhabitants show concerns mainly for municipal fragmentation and the lack of a relevant scale for investment (administrative gap), the lack of incentives for co-operation (objective gap), and the difficulty in raising tariff (funding gap).
- Compared with the other two typologies, surveyed cities with less than 1.5 million inhabitants show greater concern for the information gap. Smaller cities in the sample are more concerned with the production and the collection of data, while bigger cities are more concerned with the inconsistency of available data and their dispersion across agencies.

In terms of **demographic dynamics**:

- Surveyed cities that are growing above the OECD average are particularly concerned with the lack of relevant scale for investment (administrative gap); contradiction between levels of government and interference of lobbies (objective gap); the lack of strategic vision across water-related sectors (policy gap); the lack of staff (capacity gap); the difficulty of raising tariffs and tariff adjustments (funding gaps); and incomplete and irregular collection and over technical information (information gap).
- Surveyed cities that are growing below the OECD average show concern about municipal fragmentation (administrative gap); lack of incentives for co-operation (objective gap) and difficulties in raising tariff (funding gap).

In terms of **spatial patterns**:

- Compared to surveyed cities with a higher density, surveyed cities with lower population density are more concerned with the administrative gap; lack of incentives for co-operation (objective gap); and difficulties in raising tariffs (funding gap).
- Lower densely populated cities surveyed highlight a relatively high concern for weak stakeholder engagement; and limited information sharing and monitoring, amongst others, when compared to the answers provided by more sprawled out cities.
- Finally, lower densely populated cities are more concerned about the information gap, in particular with regard to the issue of over technical information.

Table 4.2. **Water governance gaps by types of city**

	More than 5 million inhabitants (6)	Between 1.5 million and 5 million inhabitants (15)	Less than 1.5 million inhabitants (27)	Above the OECD pop growth rate (23)	Below the OECD pop growth rate (25)	Higher density (29)	Lower density (19)	With metropolitan governance arrangements (35)	Without metropolitan governance arrangements (13)
Administrative gap									
Hydrological and administrative mismatch									
Municipal fragmentation									
Lack of relevant scale for investment									
Multiplicity of services providers									
Objective gap									
Intensive competition between local authorities									
Contradiction between levels of government									
Lack of incentives for co-operation									
Conflicts over water allocation									
Interference of lobbies									
Policy gap									
Overlapping allocation of responsibilit es									
Fragmentation of water related tasks									
Lack of strategic vision across water related sectors									
Unbalanced power between different interests									
Lack of co-ordination of legislation									
Capacity gap									
Lack of staff									
Lack of knowledge									
Poor planning									
Difficulties in doing *ex ante* evaluation									
Difficulties in *ex post* evaluation									
Funding gap									
Lack of financial guarantees for borrowing									
Limited decentralisation of fiscal power									
Golden rule									
Weak prioritisation of investment									
Lack of multi-annual strategic plans									
Scarce private sector financial contribution									
Difficulties in collecting tariffs and charges from water									

Table 4.2. **Water governance gaps by types of city** *(continued)*

	More than 5 million inhabitants (6)	Between 1.5 million and 5 million inhabitants (15)	Less than 1.5 million inhabitants (27)	Above the OECD pop growth rate (23)	Below the OECD pop growth rate (25)	Higher density (29)	Lower density (19)	With metropolitan governance arrangements (35)	Without metropolitan governance arrangements (13)
Difficulties in raising tariffs									
Affordability constraints requiring tariff adjustments									
Accountability gap									
Lack of publicly available data on drinking water quality									
Lack of publicly available data on economic and financial performance									
Lack of accounting control through regular financial audits									
Lack of benchmarking									
Lack of competitive procurement processes									
Weak judicial system for conflict resolution									
Limited information sharing									
Limited monitoring									
Weak stakeholder engagement									
Information gap									
Absence or incomplete water users' registry									
Lack of data on the water balance and quality									
Inconsistencies in available data									
Incomplete and irregular data collection									
Data dispersed across agencies									
Over technical information									
Lack of independent data									

Note: The colour system indicates the average of responses for each group of cities within the four macro-categories (size, spatial patterns, demographic dynamics, metropolitan governance arrangements): light blue = below 40% of surveyed cities consider it a challenge; blue: between 40% and 60%; dark blue: above 60%.

Source: OECD Survey on Water Governance for Future Cities (2014).

In terms of **metropolitan governance arrangements:**

- The most prominent obstacles for surveyed cities with a metropolitan governance arrangement in place are the hydrological and administrative mismatch (administrative gap); difficulties in raising tariffs (funding gap); the lack of incentives for co-operation (objective gap); the lack of staff (capacity gap); incomplete and irregular data collection (information gap), but also limited information sharing. However, since the presence of metropolitan governance arrangements might not imply prerogatives on water, a case-by-case approach is needed to understand the scope and effectiveness of such arrangements.
- Cities where no metropolitan governance arrangements took place show higher concern for the lack of relevant scale for investment (administrative gap); lack of staff (capacity gap); difficulties in raising tariffs (funding gap).

Notes

1. See Ofwat website (n.d.), http://www.ofwat.gov.uk, (accessed November 2015).
2. The Government of Singapore has been providing grants in the form of U-SAVE vouchers to help offset low-income families' bills, including water expenses. In 2013, a household staying in one- to three-room flats received an annual voucher of SGD 240 to SGD 260 (about SGD 20 to SGD 22 per month), compared to the average water bill of less than SGD 35 a month. This targeted assistance ensures that all citizens have access to affordable, high quality water for the long term. See PUB (2014), Water Pricing in Singapore, http://www.pub.gov.sg/general/Pages/WaterTariff.aspx, (accessed November 2015).
3. See World Bank (n.d.), IB-NET Database, http://database.ib-net.org/Default.aspx, (accessed November 2015).
4. "Mostly concerned" as it is taken into account the correspondent higher value of each category of gaps equal of above 60%, represented in dark blue in Table 4.2.

Bibliography

Cabrera, Jr., et al. (2011), *Manual of Best Practice, Benchmarking Water Services, Guiding Water Utilities to Excellence*, IWA Publishing.

OECD (2015a), *Water and Cities: Ensuring Sustainable Futures*, OECD Studies on Water, OECD Publishing, Paris, http://dx.doi.org/10.1787/9789264230149-en.

OECD (2015b), *Subnational Governments in OECD Countries, Key Data* (brochure), OECD, Paris, available at: www.oecd.org/regional/regional-policy/Subnational-governments-in-OECD-Countries-Key-Data-2015.pdf.

OECD (2015c), *The Governance of Water Regulators*, OECD Studies on Water, OECD Publishing, Paris, http://dx.doi.org/10.1787/9789264231092-en.

OECD (2015d), "Inventory of water governance indicators and measurement frameworks", OECD, Paris, available at: www.oecd.org/gov/regional-policy/Inventory_Indicators.pdf.

OECD (2014), "Structural reforms in France: Impact on growth and options for the future",OECD, Paris, available at: www.oecd.org/newsroom/France_StructuralReforms.pdf.

OECD (2013a), *Making Water Reform Happen in Mexico*, OECD Studies on Water, OECD Publishing, Paris, http://dx.doi.org/10.1787/9789264187894-en.

OECD (2013b), *Fiscal Federalism 2014: Making Decentralisation Work*, OECD Publishing, Paris, http://dx.doi.org/10.1787/9789264204577-en.

OECD (2012a), *Redefining "Urban": A New Way to Measure Metropolitan Areas*, OECD Publishing, Paris, http://dx.doi.org/10.1787/9789264174108-en.

OECD (2012b), *OECD Environmental Outlook to 2050: The Consequences of Inaction*, OECD Publishing, Paris, http://dx.doi.org/10.1787/9789264122246-en.

OECD (2011a), *Water Governance in OECD Countries: A Multi-level Approach*, OECD Studies on Water, OECD Publishing, Paris, http://dx.doi.org/10.1787/9789264119284-en.

OECD (2011b), *Meeting the Challenge of Financing Water and Sanitation: Tools and Approaches*, OECD Studies on Water, OECD Publishing, Paris, http://dx.doi.org/10.1787/9789264120525-en.

OECD (2010), *Pricing Water Resources and Water and Sanitation Services*, OECD Studies on Water, OECD Publishing, Paris, http://dx.doi.org/10.1787/9789264083608-en.

Ofwat website (n.d.), http://www.ofwat.gov.uk, (accessed November 2015).

PUB (2014), Water Pricing in Singapore, http://www.pub.gov.sg/general/Pages/WaterTariff.aspx, (accessed November 2015).

World Bank (2014), *IBNET Water Supply and Sanitation Blue Book 2014: The International Benchmarking Network for Water and Sanitation Utilities Databook*, http://elibrary.worldbank.org/doi/abs/10.1596/978-1-4648-0276-8_ch2, (accessed November 2015).

World Bank (n.d.), IB-NET Database, http://database.ib-net.org/Default.aspx, (accessed November 2015).

Chapter 5

Governance instruments for urban water management

Water is a fragmented sector where co-ordination is essential to manage interdependencies across multiple scales, responsible authorities and policy domains. This chapter presents and discusses a range of co-ordination mechanisms that can be employed to overcome fragmentation and identifies to what extent cities surveyed in this study use them. These mechanisms are described under the umbrella of the 3Ps framework: co-ordination across policies, places and people, with particular emphasis on rural-urban partnerships, metropolitan governance and stakeholder engagement.

Key messages

This chapter emphasises a number of interdependencies that all cities face when managing water and argues that effective governance across multiple scales, policies and authorities is a powerful means to address complexity and fragmentation through greater vertical and horizontal co-ordination across:

- water-related **policies** to avoid split incentives or contradictory measures and favour inter-sectoral complementarities, while aligning objectives and strategies and efficiently allocating resources and building capacities
- **people** involved in water-related decisions or affected by them, to raise awareness on current and future water risks, manage conflicts on water allocation and articulate interests and actions
- **places** where water is located, managed and distributed to overcome territorial mismatches across different scales and favouring co-operation between cities and their surroundings.

Consistent trends towards metropolitan arrangements are noteworthy across surveyed cities, and further use of rural-urban partnerships, contracts across levels of government, stakeholder engagement and other forms of co-ordination are also needed. For resilient urban water governance, cities need to:

- develop a strategic view of water in the broader economic, social and environmental portfolios
- co-operate with their hinterlands/rural areas, neighbouring cities, and upper levels of government
- stakeholder engagement for inclusive water governance.

The 3Ps co-ordination framework

Good water governance requires strengthening co-ordination across *policies, places and people.* Several instruments can be used to achieve the objectives of better co-ordination across these 3Ps and enhance the effectiveness, efficiency and inclusiveness of water governance (OECD, 2015a). These instruments are: vertical and horizontal co-ordination and policy complementarities; dedicated metropolitan governance arrangements, rural-urban partnership; and stakeholder engagement.

Cities are *laboratories for policy complementarities* across sectors. Local governments can respond to a variety of policy goals, from reducing water consumption to preventing disruption to the water system, in connection with other policy fields such as environment, spatial planning and energy. Hence vertical and horizontal co-ordination across levels of government are much needed for a resilient and sustainable use of the resource across users and generations. In order to overcome silos and diverging interests, a range of co-ordination mechanisms can be employed at the urban level to enhance policy complementarities.

Cities are engines of growth and pools of competitiveness and productivity. The way water is managed in cities influences the economic, social and environmental development of territories but can also generate competition across water users (i.e. households, farmers, industry). Hence greater co-ordination across *places* (urban areas and their hinterlands) supports efficiency of water use:

- Dedicated **metropolitan governance** arrangements provide a range of options to manage such interdependencies at the relevant scale through greater information-sharing, costs-saving, planning, and innovative ways of devising and implementing policies and delivering services across municipalities (i.e. multi-sectoral utilities, local public enterprises).
- A positive two-way interaction between rural and urban areas ending in rural-urban partnerships can foster efficiency in water exploitation and conservation, build synergies and align targets, manage trade-offs across categories of users and foster complementarities across places.

Being the level of government nearest to people, cities are the place where people and their needs matter. Building on this, cities should foster *stakeholder engagement* in order to create consensus and ensure the accountability of city managers and service providers to end users and citizens. Stakeholder engagement can help build trust and ownership, secure the willingness to pay for water services, raise awareness on current and future water challenges, manage conflicts on water allocation and set convergent objectives across policy areas.

Policy

Vertical and horizontal co-ordination mechanisms

A range of mechanisms help co-ordinate water policy between the city (local government), neighbouring cities and upper levels of government. More than half of surveyed cities use performance indicators to co-ordinate vertically and horizontally, share databases and information systems, or use sub-national institutions that deal specifically with water (Figure 5.1).

Figure 5.1. **Water policy co-ordination mechanisms across levels of government**

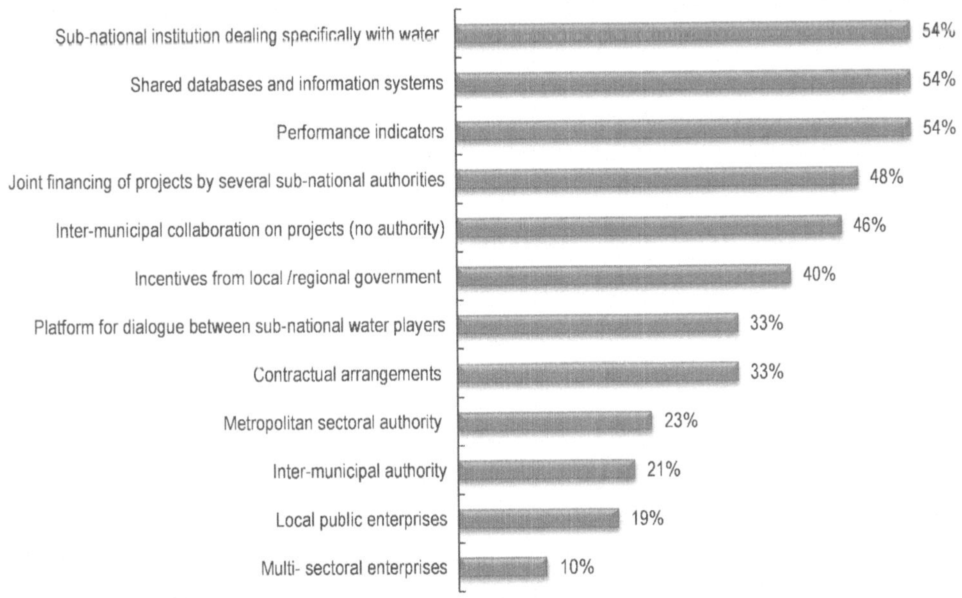

Note: Based on the responses of 48 cities, responding "yes" to the option provided.
Source: OECD Survey on Water Governance for Future Cities (2014).

Metropolitan co-ordination mechanisms can be of a different nature. When the co-ordination is favoured by formal bodies or collaboration on specific projects, involving a number of municipalities, the common goal is to reach the "critical mass" for service provision, investment, while providing effective responses to water-related issues. This is done by pooling financial and human resources, while strengthening technical skills and expertise. Examples of tools are: metropolitan sectoral authorities established within the metropolitan body (e.g. Metropolitan Water and Sewer Authority, Waterworks Authority, etc.); inter-municipal authorities (formal bodies constituted by a group of municipalities); inter-municipal collaboration on projects (involving no authority), e.g. to meet the costs and requirements for the construction, operation and maintenance of water-related infrastructure (Box 5.1).

Box 5.1. **A metropolitan steering committee to cope with water scarcity in Bologna**

In order to ensure co-ordination of the actions necessary to address the summer droughts that affect the metropolitan area of Bologna, a steering committee (*cabina di regia*) involving organisations with expertise in water management was set up in 2013. The steering committee is composed by Emilia-Romagna (president of the steering committee), the Territorial Agency of Emilia-Romagna for Water Services and Waste (ATERSIR), the Province of Bologna, the Water Service Operator HERA Bologna, a number of consortia and municipalities.

The steering committee meets four to five times a year to agree on measures to monitor the effective implementation of water policies and to provide updates when needed. The steering committee can play an important role in the implementation of the Adaptation Plan for Climate Change issued by the City of Bologna. Next steps are to extend the competences of the steering committee to flood risk management and to increase the frequency of meetings to address other water risks.

Source: Inputs provided by the city of Bologna, Comune di Bologna (n.d.), "BLUE AP Project, Strategia di adattamento ai cambiamenti climatici della Città di Bologna", brochure, available at: www.blueap.eu/site/wp-content/uploads/2013/09/ BLUEAP_Strategia_adattamento_locale.pdf.

A number of mechanisms can be used at the basin level, which is usually the scale at which it is recommended to integrate physical, environmental, social and economic aspects on water resources management. Sub-national institutions dealing specifically with water, such as RBOs and water agencies, therefore, contribute to this integrated basin governance while co-ordinating water policy across sub-national actors, between levels of government and engaging with stakeholders.

Different mechanisms can also be used for organising service provision as they allow for the co-ordination within a certain number of jurisdictions to secure economies of scale and promote complementarities. Multi-sectoral enterprises and local public enterprises are two examples of management models for service delivery, which can bundle more services together, such as water, energy and waste (Saussier and Klein, 2014).

A number of tools can also foster dialogue across authorities and levels of government. This is the case of technical committees, ad hoc negotiating tables as occurs in the City of Montreal, but also contractual arrangements clarifying responsibilities and enabling interactions, while building capacity across local authorities (OECD, 2007; Charbit and Romano, forthcoming). A good practice is the EPA Water Infrastructure and Resiliency Finance Center set up in April 2015 to help US municipalities efficiently use federal and local funds for water infrastructure, explore financing options and showcase best practices[1].

Horizontal and vertical co-ordination can be promoted through incentives, be they monetary or non-monetary, from local/regional governments. Such incentives might consist in rules, rewards, sanction mechanisms, earmarked funding, or joint financing of projects by several sub-national authorities to pool resources and capacity at the relevant scale.

Shared databases and information systems can improve the knowledge base and push governments to collaborate towards better decision making when information is reliable and robust. Performance indicators help assess the performance of the sector in terms of outcomes in most cases. When measuring progress, performance indicators are employed for monitoring more than for evaluation, which in turn establishes causal relationship between inputs and outputs (OECD, 2009 OECD, 2011). In both cases, they can help to identify areas where co-ordination can be improved and support negotiation for better allocation of resources or competences.

Policy complementarities

Policy coherence across sectors is crucial, as regional development, land management, agriculture and energy policies, amongst others, affect water in cities (Figure 5.2). In addition, water outcomes are often driven by decisions made in policy areas over which water managers have little or no say (OECD, 2011).

Figure 5.2. **Policy areas influencing water challenges in surveyed cities**

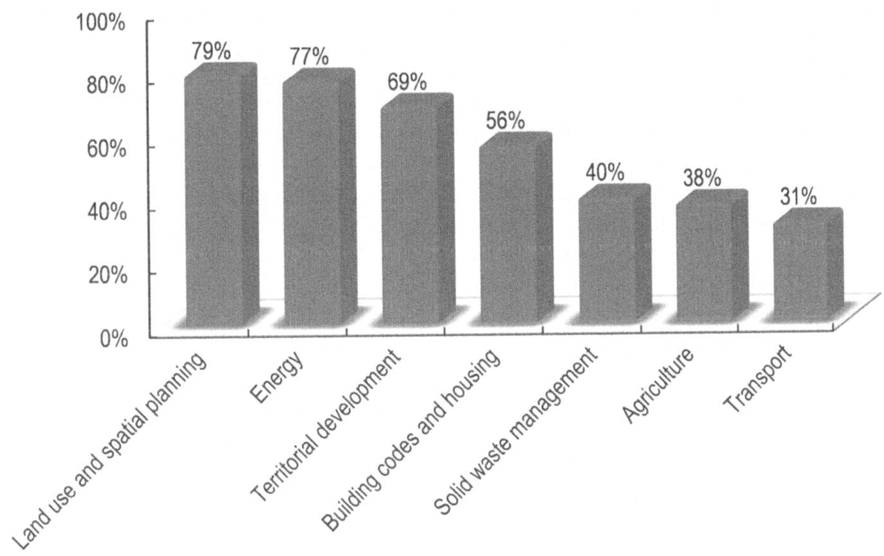

Note: Results based on 48 responses that indicated the influence from policy areas being "critical" and "important".

Source: OECD Survey on Water Governance for Future Cities (2014).

Land use and spatial planning influence the way water is managed within the city's boundaries and can contribute to integrated water resources management and water security according to 79% of the cities surveyed. Land use planning is linked to the development of zoning plans or land use plans, which directly affect property rights and landowners. Spatial planning includes different spatial levels (national, regional, and municipal planning). The plans are usually strategic plans (i.e. the Dutch *structuurvisie* or the German *Flächennutzungsplan*). There are many important points that make this relevant: the spatial scale, participation issues, integration of sectoral policies (i.e. usually strategic spatial planning is more integrative, land use planning more pragmatic).

Urban flooding in inadequately maintained built-up areas requires measures to manage run-off, which increases with impervious surfaces (roads, buildings) that do not absorb surface water. Run-off generated in areas of urban sprawl is approximately ten times greater than that in more dense urban areas. Specifically, large parking lots and wider roads often cause higher levels of run-off (OECD, 2014a). This calls for more systematic and integrated policy packages between water and urban planning. The City of Cologne, for example, co-ordinates water and spatial planning for new building areas to prevent flood damages because of heavy rainfalls. Municipal authorities in the Netherlands, through the "Water Assessment" instrument, take water management into account in their spatial planning decisions. Even though it is not binding, this tool is considered effective for linking water authorities and cities (OECD, 2014b).

Energy and water are strictly related for both energy production and for getting water from alternative sources, and such interdependency was considered among the highest rates of linkages by 77% of cities surveyed. Energy production is strongly dependent on water for power generation, fossil fuel extraction, transport and processing, and irrigation of biofuels crops. The energy sector accounts for an estimated 15% of global freshwater use, mostly for cooling at thermal power plants. It is second only to agriculture and it can pose risks to water quality (IEA, 2012). The development of non-fossil fuel energy sources, such as hydropower and biofuels, has put serious pressure on water resources. Water scarcity may force the closure of power plants that require freshwater for cooling (OECD, 2011).

Increasing efficiency of water-energy cycle is an important goal for water managers, especially within the context of climate change. Electricity is a heavy feature on the annual budget of water utilities, due to operations such as pumping from withdrawal and to treatment plants, which often occur outside the populated areas. A number of good practices exist to foster greater coherence among water and energy policies at the local level. In Budapest, for example, legal requirements are used for co-ordination between water utility supply and energy sectors. To cope with water scarcity, alternative water sources (such as desalination and reuse) have been developed, but they pose challenges on energy consumption. Amongst the cities surveyed, Barcelona, Hong Kong, China and Singapore (Box 5.2), have been investing in alternative sources of water.

Territorial development is the third area closely related to urban water management according to 69% of surveyed cities. On the one hand, water is essentially a local issue (pumped, treated, distributed and used locally), but on the other hand, it has implications beyond the local scale as water drives or hinders economic development, competitiveness and assets of regions in a national framework. Given the importance of local actors and territorial specificities in the water sector, policy makers should find ways to maintain coherence while preserving diversity.

Box 5.2. Singapore co-ordination strategy: Water-energy-waste

Energy consumption is and will continue to be a challenge in Singapore's water supply and used water operations. The Public Utilities Board (PUB), Singapore's national water agency, is continuously finding ways to mitigate the impact of energy on the processes. Singapore has put in place a long-term water supply strategy known as the "Four National Taps": i) local catchment water from the reservoirs; ii) imported water from Malaysia; iii) NEWater: ultra-clean, high-grade reclaimed water; iv) desalinated water.

Amongst the Four National Taps, desalination is the one with the highest energy consumption. With the aim of at least halving the current energy levels, PUB has partnered with Evoqua Technologies (previously Siemens Water Technologies Corporation) to pilot electrically driven processes to desalt seawater and other innovations. PUB is looking into building rooftop solar panels at waterworks and installing floating solar systems on the reservoir to explore alternative and sustainable energy sources.

There is a plan to co-locate the Tuas Water Reclamation Plant (TWRP) by 2017, which will incorporate technologies to improve energy efficiency and manpower requirements, with the National Environment Agency's Integrated Waste Management Facility, to reap the potential synergies of the water-energy-waste nexus. This co-location marks Singapore's first initiative to integrate used water and solid waste treatment processes to maximise both energy and resource recovery, while minimising land footprint.

Source: Inputs provided by the City of Singapore; OECD Survey on Water Governance for Future Cities (2014).

Building codes and housing are increasingly aiming at reducing water consumption and protecting from water-related risks; they represent an area of interdependency for 56% of cities surveyed. Green buildings are those with increased energy efficiency and reduced water consumption, since the less water is used for related end uses such as heating, cooling and preparation of hot sanitary water, the less energy is required. Water-related activities in buildings, in fact, require approximately 20-25% of domestic energy consumption (IEA, 2009). Trends show that water demand for domestic use is projected to increase by 130% by 2050 (OECD, 2012a). Good options for significant water savings are recycling rainwater and greywater, but quality standards of reused water for non-potable domestic uses in building need to be carefully set to avoid health-related issues, which are the major constraint in the development of such systems (OECD, 2012a).

Solid waste management can impair quality and quantity when discharged into surface water, and was reported as an important field for policy coherence by 40% of surveyed cities. Poor waste collection practices and improper municipal solid waste management can contribute to water resource pollution, in terms of surface and groundwater contamination. While this issue is more alarming in developing countries, developed countries still face problems of illegal dumping, especially in the case of industrial and special waste (e.g. chemicals). The City of Milan is combining water resources management and waste management under the umbrella of the project "Smart Water Resource Management".[2]

Agriculture also raises urban water governance challenges, for example in terms of controlling and avoiding the harmful consequences of the use of fertilisers and pesticides on water. Agriculture is the largest consumer of water and an important source of water pollution, especially in the case of intensive agricultural practices, using excessive inputs of nutrients, nitrates and phosphates from fertilisers (OECD, 2012b). Support for agricultural production and some subsidies (e.g. to agricultural inputs) contribute to water overuse and pollution. Projections over the next ten years foresee an intensification of agricultural production with negative consequences on water systems in some countries (OECD, 2012b). The impacts of agricultural use are ubiquitous, but often vary from place to place. In some places it is linked to fertilisation and pesticides, in others it may be linked to over abstraction or to excessive drainage (e.g. urban flood risk linked to disappearance of floodplains). In 2014, Eau de Paris put in place an information campaign and concrete actions to promote organic agriculture for the preservation of water and natural resources.[3] There are also contracts between the municipality of Paris, authorities in the hinterland and farmers for developing a tighter co-operation between supplying areas in terms of water resources and the urban core (see Box 5.9 in the section on stakeholder engagement).

Transports have both direct and indirect impacts on water quality and typically affect water in cities for run-off management, while impervious surface can alter the movement of water in urban areas. New York's "Green Infrastructure Program" promotes the resilience of the drainage system by intercepting stormwater run-off and enabling it to infiltrate the ground rather than entering the sewers.

Cities use a range of mechanisms to co-ordinate water and related policies (Figure 5.3). Planning instruments are the ones most commonly used across sectors, especially for articulating water and land (58%). Planning is followed by co-ordination groups or meetings, and regulations which are respectively more prominent for water and the environment (52%) and for water and land (44%). Cities also resort to contracts to co-ordinate water and energy (27%), for example to co-ordinate between water utility supply and energy sectors. Contracts are also employed for water and agriculture as experiences of decentralised urban-rural arrangements aim to reduce water pollution from agriculture, while agreements with farmers try to reduce flood risk upstream.

Figure 5.3. **Co-ordination between water and related sectors**

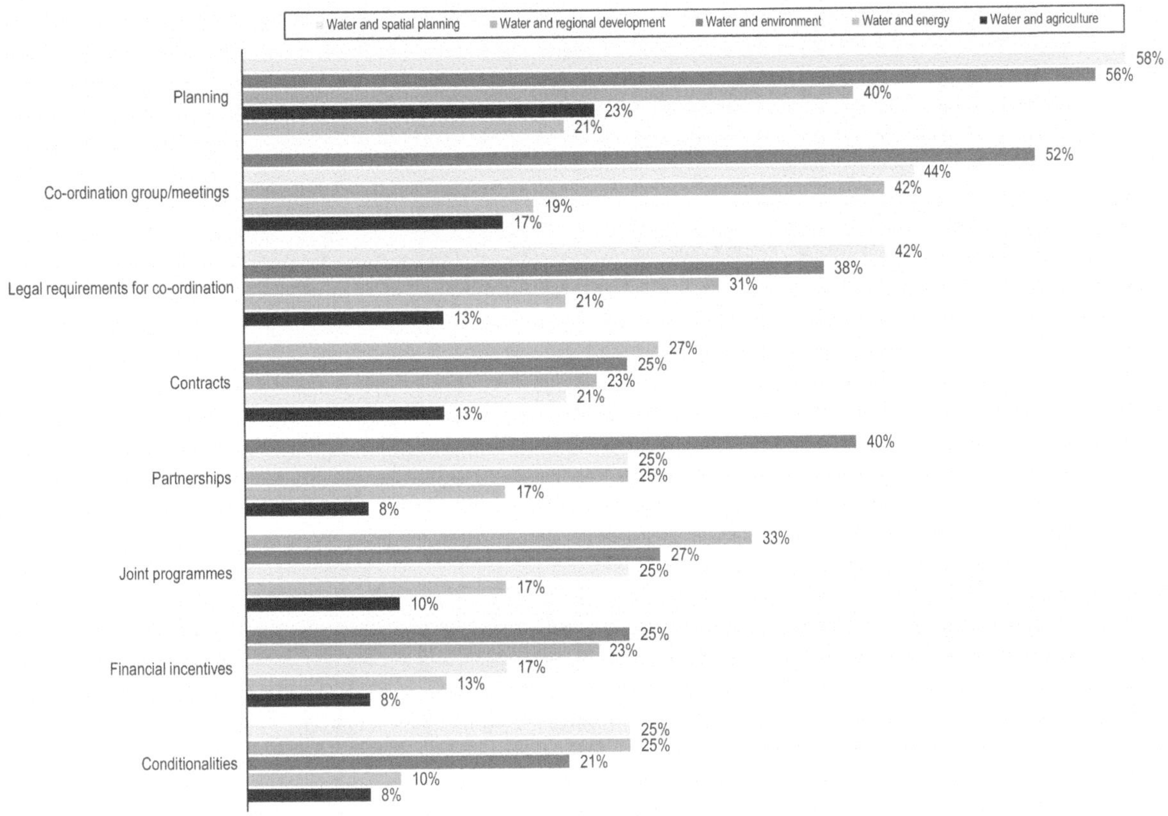

Note: Results based on 48 responses that indicated co-ordination of water management with other sectors.

Source: OECD Survey on Water Governance for Future Cities (2014).

Places

Metropolitan governance

OECD (2014c) pointed out that building effective metropolitan governance for stronger, more inclusive and sustainable growth is all the more salient in a context of economic crises and pressure on public finances. Metropolitan areas face several issues concerning the water sector: the articulation between institutional, functional and hydrological logics, the problem of the appropriate scale of water resources management and service delivery; water conflicts between municipalities within metropolitan areas silo approaches in the absence of multi-level governance.

OECD (2015b) investigated four governance arrangements, which can also have an impact on the management of water between the urban core and the surroundings, in terms of investment needed, sharing information, monitoring, stakeholder engagement, policy complementarities across different sectors. They are summarised in Figure 5.4 and concrete examples of their use in surveyed cities are provided in the next sections of this chapter. Table 5.1 provides examples of metropolitan governance arrangements in place in surveyed cities. These arrangements do not necessarily imply competences on the water sector.

Figure 5.4. **Options for metropolitan governance arrangements and selected examples on water**

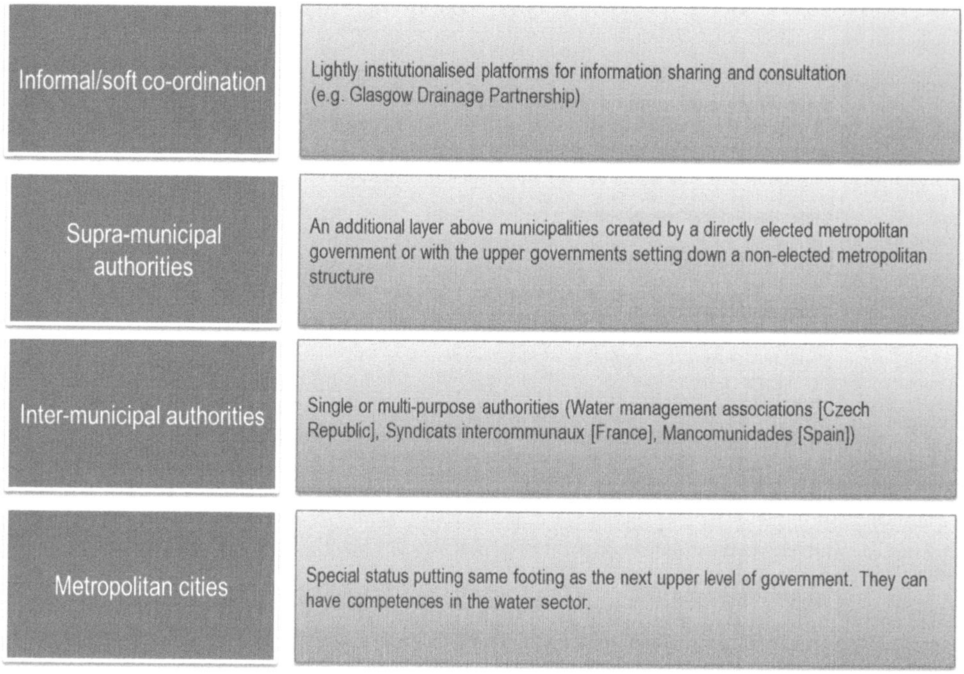

Source: Based on OECD (2015b), *Governing the City*, OECD Publishing, Paris, http://dx.doi.org/10.1787/9789264226500-en.

Table 5.1. **Metropolitan governance arrangements in place in surveyed cities**

	Metropolitan governance arrangement (not water-related)
Acapulco de Juarez	Consejo para el Desarrollo Metropolitano del Municipio de Acapulco (CODEME)
Amsterdam	Stadsregio Amsterdam
Barcelona	Àrea Metropolitana de Barcelona
Belo Horizonte	Belo Horizonte Metropolitan Region
Bologna	Città Metropolitana di Bologna
Budapest	Budapest Metropolitan Development Council
Calgary	Calgary Regional Partnership
Chihuahua	Consejo para el Desarrollo Metropolitano de Chihuahua
Cologne	Region Köln-Bonn e.V.
Copenhagen	Capital Region of Denmark
Culiacan	Comisión de Zona Conurbada de los Municipios de Culiacan y Navolato
Daegu	Daegu Metropolitan City
Edinburgh	Strategic Development Planning Authority for Edinburgh and South Scotland
Glasgow	Glasgow and the Clyde Valley Strategic Development Planning Authority (GCVSDPA)
Grenoble	Grenoble-Alpes Métropole
Hermosillo	Consejo para el Desarrollo de la Zona Metropolitana de Hermosillo
Kitakyushu	Kitakyushu Metropolitan Area Promotion Council
Lisbon	Area Metropolitana de Lisboa
Marseille	Marseille Provence Métropole
Mexico City	Consejo para el Desarrollo Metropolitano del Valle de México

Table 5.1. Metropolitan governance arrangements in place in surveyed cities *(continued)*

	Metropolitan governance arrangement (not water-related)
Milan	Città Metropolitana di Milano
Montreal	Communauté Metropolitaine de Montreal (CMM)
Nantes	Nantes Métropole
Naples	Città Metropolitana di Napoli
Paris	Paris Metropole
Phoenix	Maricopa Association of Governments
Queretaro	Consejo Para el Desarrollo Metropolitano de la Zona Metropolitana Ciudad de Queretaro
Rome	Città Metropolitana di Roma Capitale
San Luis Potosi	Consejo para el Desarrollo Metropolitano del Estado
Stockholm	Kommunförbundet Stockholms Län
Toluca	Consejo para el Desarrollo Metropolitano del Valle de Toluca
Turin	Città Metropolitana di Torino
Tuxtla	Consejo de Desarrollo Metropolitano de Tuxtla Gutierrez
Veracruz	Consejo para el Desarrollo Metropolitano del Estado de Veracruz

Source: Ahrend, R., C. Gamper and A. Schumann (2014), "The OECD Metropolitan Governance Survey: A Quantitative Description of Governance Structures in large Urban Agglomerations", *OECD Regional Development Working Papers*, No. 2014/04, OECD Publishing, Paris, http://dx.doi.org/ 10.1787/5jz43zldh08p-en;.OECD Survey on Water Governance for Future Cities (2014).

The Metropolitan Glasgow Strategic Drainage Partnership (MGSDP) is an example of an informal/soft co-ordination arrangement. It is a collaborative venture between local authorities (Glasgow City Council leading), the Scottish Environment Protection Agency (SEPA), Scottish Water, Scottish Enterprise, Clyde Gateway and Scottish Canals. It was set to upgrade the Glasgow area's drainage and sewerage network, reduce flooding and support urban development requirements, and improve water quality and the environment. In more than ten years of activity the partnership has produced tangible benefits, such as the reduction of risk of flooding to 7 000 properties and GBP 40 million investment to improve water quality.[4]

There are also several examples of inter-municipal authorities:

- The governing bodies of the optimal territorial areas (*Ambito Territoriale Ottimale*) in Italy are autonomous entities defined by the Region and made up of municipalities. They are responsible for planning, defining investment programmes and selecting service providers.[5] They usually cover an area correspondent to the province (Box 5.3).
- The Water Management Association of the West Bohemia Region in the Czech Republic is a voluntary union of 93 municipalities and two associations of municipalities from five districts (Karlovy Vary, Tachov, Sokolov, Chomutov and Rakovník). The association carries out the management, operation and development of water supply and wastewater treatment systems. The governing body of the association is the General Meeting, where each municipality has one vote.[6]

Box 5.3. **The Authority of the optimal territorial area in Turin**

The authority of the optimal territorial area (*Ambito Territoriale Ottimale,* ATO), *Autorità d'ambito n. 3 "Torinese"*, encompasses a large area covering 306 municipalities, located in the Metropolitan area of Turin. The authority is defined by the Region, which establishes also the modalities of participation of local entities within the ATO. The authority manages water services in an integrated way and defines water-related infrastructure planning.

The Authority carries out various functions:

- specifies the level and the quality of the Integrated Water Management Service to be provided to the users
- adopts an implementation programme for infrastructure and equipment for service delivery
- set ups the financial plan and the tariffs level
- set ups the management model and investigate on modalities of production of the Integrated Water Management Service
- carries out the necessary legal action for outsourcing the water service
- ensures the operational, technical management and monitoring
- investigates recharging sources.

Source: Input provided by the City of Turin, Autorità d'ambito Torinese website (n.d.), www.ato3torinese.it/loader.html?_organigramma/conferenza.html (accessed December 2015).

- The *Conseil communautaire* (France) is an elected body that can act on behalf of the municipalities on specific water issues (i.e. water allocation; drinking water provision; research; operation/maintenance of infrastructure). The *Syndicats intercommunaux* are run by joint committees representing members of each local council and levying a compulsory contribution for water supply.
- *Mancomunidades* (Spain) are administrative forms meant for purely inter-municipal co-operation in which municipalities appoint local politicians for the governing body of the *mancomunidades*, the number of which is proportional to the size of the population of the respective member municipalities. They help water services reach economies of scale (OECD, 1999).

Another kind of co-operation based on the participation of higher levels of government are *Consortia* (Italy, Spain), which are standing organisations with a board and staff for drinking water supply cycle (from production to distribution) (i.e. Greater Bilbao Water Partnership, a consortium of 43 municipalities, provincial government of Biscay, the Autonomous Basque Community and central government; OECD, 2015b).

More than half of the sample of surveyed cities reported the existence of a dedicated metropolitan governance body covering their administrative boundaries, but it does not necessarily have prerogatives over water management (Box 5.4). Out of 27 cities with a metropolitan body, 23 reported a number of water-related competences, as indicated in Figure 5.5. Digging deeper on the responsibilities on water competences of metropolitan bodies, the OECD Survey on Water Governance for Future Cities shows that they mainly act as policy facilitator, favouring information exchange across the municipalities in the metropolitan area (Figure 5.5). Metropolitan bodies also tend to provide technical expertise, strategic management of service provision (e.g. setting of performance targets, hiring of senior managers, call for tenders or supervision of sub-contractors) and use legislative, regulatory or other authoritative competences.

Box 5.4. **Examples of metropolitan water governance in Barcelona and Nantes**

Barcelona Metropolitan Area

The metropolitan area of Barcelona is formed by 36 municipalities of which Barcelona is the largest. From a hydrological point of view, the 36 municipalities are managed as a unique territory by the Metropolitan Authority (AMB). The AMB has regulatory and statutory authority. In the water sector the AMB approves regulations governing the integrated water cycle (supply, purification, distribution and sanitation) and the discharge of wastewater to the metropolitan sewage. It also approves tax ordinance to regulate the fees connected with the services and carries out administrative activities. The AMB has jurisdiction over most of these activities, which promotes integrated management of water supply and sanitation in the metropolitan area.

In the metropolitan area, there are seven wastewater treatment plants and three reclaimed water plants. Managing urban waters at the metropolitan level has fostered an integrated perspective of the water cycle level, as well as shared infrastructure and expenses. The AMB encourages customers' involvement to learn about different territorial needs and expectations. Next steps for AMB include looking at alternative sources for water and strengthening the water cycle management. Aguas de Barcelona was created in 2013, jointly with a large metropolitan utility, to manage drinking, reclaimed and wastewater for all the metropolitan territory.

Nantes Métropole

Nantes Métropole has prerogatives over the water policy in the 24 municipalities of its territory. This policy covers the entire water cycle: drinking water and collective sewage and wastewater, but also restoration of aquatic environments and storm water management. Since 2001, Nantes Métropole has been managing infrastructure that supplies drinking water and sanitation for nearly 600 000 inhabitants and chose the mix of management models in its territory. Nantes Métropole is in charge of the organisation of general public water services (collective and non-collective), and has the title of Organising Authority that sets the level of tariffs, defines the pricing policy for water and sanitation, and evaluates operators' performance.

Source: Input provided by the Area Metropolitan Barcelona and Nantes Métropole to OECD Survey on Water Governance for Future Cities (2014) and Àrea metropolitana de Barcelona website (n.d.), www.amb.cat, (accessed October 2015); Marest, P. et al. (2006), "Sustainable water management, the choices of Nantes Métropole", In Urban Public Services: the triptych organising authority, operators, citizen-users, Local powers n°71 III/ December 2006.

Figure 5.5. **Key functions of metropolitan bodies with water responsibilities in surveyed cities**

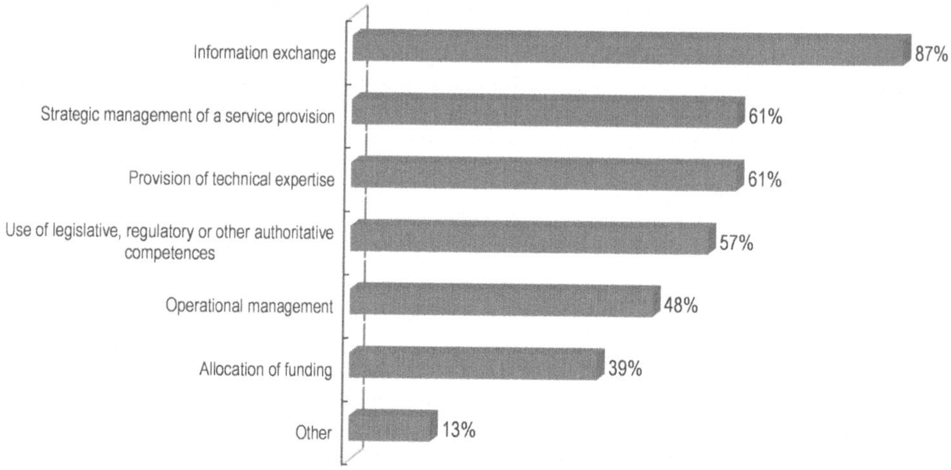

Note: Based on the responses of 23 cities indicating the existence of metropolitan bodies with water competences.

Source: OECD Survey on Water Governance for Future Cities (2014).

Strategies for water governance at the metropolitan level may offer interesting models for application in the sector: these governance arrangements are likely to affect water management between the urban core and the surroundings in terms of investment, information sharing, monitoring, stakeholder engagement, policy complementarities across different sectors. Tangible projects, as those concerning public services, can motivate the collaboration at metropolitan level and represent a first step for effective metropolitan governance reforms (Box 5.5).

Box 5.5. **Steps for effective metropolitan governance reforms**

Drawing from case studies the OECD suggests five steps for effective metropolitan governance reforms:

1. **Motivate collaboration by identifying concrete metropolitan projects**. This step consists in identifying the framework conditions as favourable (economic, social and political ones. For example a metropolitan governance reform is facilitated when authorities have a clear electoral mandate for change), start up the collaboration on tangible projects, as those concerning public services.
2. **Build metropolitan ownership among key stakeholders**. Strong political will is a crucial factor, but a drive for reform may derive not only from the municipal level itself, but from the central government, or also other local sub-national governments. Also citizens and communities need to be involved since the beginning for a stronger legitimacy of the reform.
3. **Tailor reliable sources of metropolitan financing**. Metropolitan reforms should define how the new governance structure can help respond to the financial needs of the metropolitan region, and how to match the new governance structure's responsibilities with corresponding financial resources among the main sources of revenues.

Box 5.5. **Steps for effective metropolitan governance reforms** *(continued)*

4. **Design incentives and compensations for metropolitan compromises**. OECD experience suggests that co-operation among municipalities works best on a voluntary basis with incentives from the top, but also when a strategy is elaborated for engaging those who feel threatened by the reform and leveraging their buy-in. It is important to communicate the benefits of reforms and the costs of non-reform. Incentives may consist in giving a range of new powers to cities.
5. **Implement a long-term process of metropolitan monitoring and evaluation**. Solid background research and scrutiny from unbiased experts can help create and sustain credibility for the reform by strengthening the evidence base. Independent expertise and research capacity are required to demonstrate the need for change and the desirability of the proposed solutions to key stakeholders, as well as analyse and weigh different options against each other.

Source: Adapted from OECD (2015b), *Governing the City*, OECD Publishing, Paris, http://dx.doi.org/10.1787/9789264226500-en.

Rural-urban partnerships

Urban and rural areas are subject to many interdependencies related to the risk of *too little water* (the projected growth in competition for water resources, linked in particular to rising uncertainties about water supply due to climate change), *too much water* (mitigation of flood risks, which are bound to increase in the future) and *too polluted water* (the maintenance of adequate quality standards for water, as a source of drinking water and for the provision of ecosystem services) (OECD, 2015c). According to the results of the Survey on Water Governance for Future Cities, these interdependences are related to water quality (75%), quantity (73% flood control mechanisms), infrastructure (63% wastewater treatment) and water allocation (58%), amongst others (Figure 5.6). The use of water in urban cores reduces the availability for agriculture, ecosystems and regional development in rural areas and vice versa. In addition, the issue of water quality creates interdependences in terms of point and non-point sources of pollution from agricultural and from urban run-off, with consequences on both areas. Rural areas might also suffer from the discharging of urban wastewater or they can benefit from policies from wastewater reuse for agricultural irrigation, as an extra source of water available for the rural sector.

Figure 5.6. **Issue generating interdependencies between cities and surrounding areas**

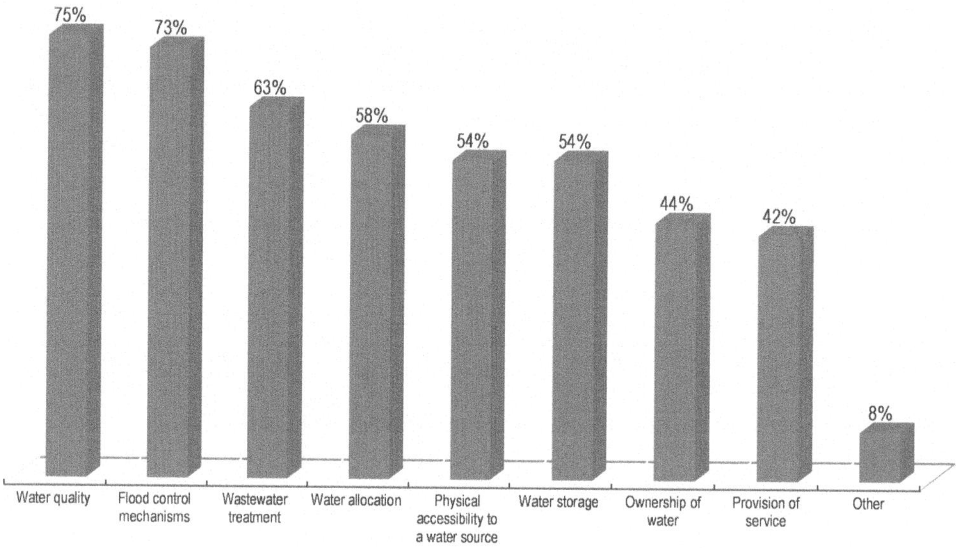

Note: Results based on a sample of 48 respondents who responded "yes" to the options provided.

Source: OECD Survey on Water Governance for Future Cities (2014).

Urban areas depend on their surroundings for getting water, providing water and protecting their citizens from extreme water-related events. Rural areas also represent an asset for the sustainability and resilience of expanding urban environments, but in some cases they lack improved water sources. The lack of proper connection with wastewater treatment infrastructure can generate significant health and social concerns at the local and national levels (OECD, 2012a; OECD, 2012b). The interaction between urban and rural areas should promote a more balanced use of water, the use of new sources of water (i.e. recycling urban wastewater), the upgrade in technology for reducing the consumption of energy and water. Boosting water efficiency also requires technological upgrade for irrigation practices, the use of alternative sources of water, smart systems for reducing water consumption and green infrastructure for flood prevention.

To match the projected demand of the increasing population, world agricultural production would need to increase by some 60% between 2005 and 2050 (Alexandratos and Bruinsma, 2012). This would have to be achieved with less water, mainly because of pressure from growing urbanisation, industrialisation and possibly climate change. This implies an urgent need to adopt water-efficient irrigation technologies, such as drip emitters, and better maintenance of irrigation infrastructure. Linked to food production is the issue of food waste: FAO 2011 states that globally about one-third of food produced for consumption is lost or wasted each year, implying that resources, including water, are used in vain. Governance then becomes key to raise awareness, catalyse the financing needed for these investments, to ensure stakeholder buy-in, to share data across institutions and places, and to strengthen policy coherence between water, land use, energy and agriculture. Examples of urban-rural interconnection across surveyed cities are presented in the Box 5.6.

Box 5.6. Examples of rural-urban co-operation in cities

Milan: Enhancing the rural area of Milan through the plant Milan Nosedo

The City of Milan is located in the Po valley, known for the fertility of its land and its water wealth. The purification plant for the City of Milan, Nosedo, can be seen not only as a site designed to meet the cleansing needs of the City of Milan, but as a real centre of "urban experiment", which combines social welfare, rural development, land management, renewable energy, social and environmental education. The plant produces reusable water for agriculture, serving more than 90 farms. In order to complete its work of rehabilitation and enhancement of the agricultural tradition of the territory, the City of Milan started the construction of a large public park called "Park Vettabbia", where the plant is located. In particular, the design of the new Park Vettabbia defines the new structure of waterways system, giving to the entire system a high environmental value.

Montreal: A multi-stakeholder committee to improve the quality of discharged water in catchment areas

In the metropolitan area of Montreal, every city is confronted with the challenge of collecting, treating and discharging water while also maintaining good water quality in catchment areas. On a daily basis, cities' technical services around Montreal are carried out for the maintenance and development of wastewater collection networks while the City of Montreal manages treatment facilities before water is discharged to the river. A technical committee composed of representatives from community organisations, the industrial sector, government departments, other levels of government and municipal services, was created to centralise and share information related to wastewater discharge and water quality, and to target priority areas. The collaboration between these bodies has ensured the gradual improvement of water in catchment areas.

The New York City watershed programme

The NYC water supply watershed lies outside of New York City. It is largely forested, with agricultural, suburban, and rural community activities. Policy and regulation enforcement are conducted by the Department of Environmental Protection (DEP) in order to protect source waters for use by consumers in the city and in certain upstate counties. There are also numerous release regulations and policies, as well as stakeholder meetings and co-ordination groups. Watershed protection initiatives, some of which involve the DEP and other partners, include such measures as the Watershed Agricultural Program (WAP), the acquisition of watershed lands, and the enforcement of updated Watershed Rules and Regulations.

In 1997 New York City, the watershed communities, the US EPA, the State of New York, and environmental organisations signed a Memorandum of Agreement to preserve both water quality and the economic dynamism of watershed communities. The agreement established the institutional framework and relationships needed to implement the range of protection programmes identified as necessary by the City, State and EPA.

Paris: Programmes and contracts to protect catchments in suburban and rural areas

The city water operator, Eau de Paris, has been involved in two programmes – Phyt'Eaux Cités and Preri - to preserve and improve water quality in its catchment areas, in partnership with the river basin agency of Seine-Normandie. The first programme, Phyt'Eaux Cités, encourages suburban communities, golf courses, garden centres and transportation networks to reduce or stop their use of pesticides in the Yvette, Orge and Seine basins. The second programme, Preri, aims to prevent industrial risks near the Seine and Yerres rivers by identifying and monitoring potentially dangerous sites in terms of industrial waste.

Eau de Paris also assists farmers in developing sustainable agricultural approaches that take water quality restoration and protection into account. More than 140 farmers have committed, alongside Eau de Paris, to significantly reduce the amount of nitrates and pesticides in their crops. Most recently in June 2015, Eau de Paris adhered to the charter of the National Federation of Organic Agriculture (FNAB), and thus remains committed to develop organic agriculture in its rural catchment areas. Since 2009, these efforts have helped multiply by six the size of organic farming lands in the catchments areas of the Vanne valley.

Source: Inputs provided by the cities of Milan, New York, Paris to OECD Survey on Water Governance for Future Cities (2014); Depuratore Nosedo website (n.d.), www.depuratorenosedo.eu, (accessed October 2015); City of Montreal (2007), *Le Réseau Bleu – Montréal vers une culture de l'eau*, Reference Framework, July; Eau de Paris website (n.d.), www.eaudeparis.fr/ (accessed September 2015); Watershed Agricultural Council website (n.d.), www.nycwatershed.org, (accessed September 2015).

Managing rural-urban interdependences requires co-ordination for a win-win approach between cities and their surroundings. Rural-urban partnerships are mechanisms for co-operation across rural and urban areas, but differently from other forms of co-operation, both urban and rural areas must be directly involved in the process (e.g. urban and rural municipalities; private agents [firms, civil society, etc.]). The partnership is based on a common set of objectives intended to be managed jointly, in a space where urban and rural dimensions are physically and/or functionally integrated (OECD, 2013).

Rural-urban partnerships consist in cross-sectoral and holistic sets of initiatives (e.g. within a wider package of environmental policy initiatives) or are focused on single objectives/projects (i.e. management of water resources). In Forli-Cesena (Italy) for example, the management of water resources is the result of a partnership among all urban/rural municipalities and chambers of commerce from three different provinces, which are also included in the co-operation process. Municipalities where the water sources are located benefit from a share in the revenues from water provision, as well as from investments in natural and cultural heritage preservation and initiatives aimed at developing tourism in the area. At the same time, the other municipalities benefit from the availability of clean water and from the proximity to high-value landscape and amenities (OECD, 2013).

Successful rural-urban partnerships require mutual trust and clear understanding of the long-term benefits of the interaction. Differences between rural and urban areas in terms of capacity, economic and political power can complicate this relationship (OECD, 2013). In economic and social terms, there is an imbalance of power between urban and rural, where urban often definitely prevails. The imbalance can stem from the lack of information, evidence, and data as well as the lack of capacity in rural areas. However, this is not always the case in the water sector, where water management in rural areas for agricultural production generates negative externalities on urban dwellers in terms on water pollution or increased costs (e.g. in southeast Spain urban dwellers have to pay for desalinated water as "cheap" subsided water is used for agricultural purposes).

The institutional framework (regulatory and political barriers) can sometimes constrain rural-urban partnerships. The absence of proper mechanisms or incentives for co-operation can also undermine a rural-urban partnership even when there is interest on both sides. Rural-urban co-operation is not always risk-free: it can entail excessive transaction costs and additional administrative burdens. The process of co-ordination across levels of government, in fact, requires time and resources, as well as capacities that may be not in place at the start of the co-operation process (OECD, 2013). Given this complexity, partnerships should be based on careful considerations of costs and benefits of this co-ordination mechanism, as well as of the timeframe of action and the complexity of modelling that is required to actually bring information from the hinterland into the city solutions (Box 5.7).

Box 5.7. Conditions for the success of rural-urban partnerships

Two types of rural-urban partnerships can be distinguished:

- **Explicit**: If the rural-urban dimension is very clear and the linkages between the urban and rural stakeholders are taken into account in the partnership's membership mix, work and strategic objective.
- **Implicit:** If the partnership aims to improve co-operation through a common local development objective, strategy or project but still involves urban and rural authorities.

There are three different designs that can shape the form of co-operation:

- **Formal and institutionalised**: A formal commitment by the actors involved to reach out across their respective responsibilities and interests and to co-operate on certain issues.
- **Formal but not institutionalised**: Hybrid partnership with features of both formality and informality. It has all the characteristics of the first group except that it is not institutionalised and is less structured and looser. It has no independent structure, with staff or allocated resources.
- **Informal**: Members decide to join together in loose networks that permit mutual consultation and co-ordination. No particular body is laid down, rules for co-operation are not well developed, and competences are limited.

A successful rural urban partnership for water management relies on five pillars, for which the role of local government is crucial.

- **Promote a better understanding of socio-economic conditions in urban and rural areas**. The local government can use the rural-urban governance framework to promote water quality, reduction in water consumption and flood protection.
- **Address territorial challenges with an approach based on functional linkages between urban and rural areas**. The local government can identify strengths and weaknesses of rural-urban areas in managing water and their interdependencies.
- **Encourage the integration of urban and rural policies by working towards a common agenda**. The local government can encourage the participation of different government levels in rural-urban partnerships to achieve better policy integration as well as aligning interests inside and outside the water box.
- **Promote an enabling environment for rural-urban partnerships**. The local government can develop trust and a shared vision of the territory by promoting pilot projects on easy "win-win" issues, education initiatives, dialogue facilitators and setting a balanced "rules of the game" (i.e. promoting a fair partitioning of voting rights within the partnership).
- **Clarify the partnership objectives and related measures to improve learning and facilitate the participation of key urban and rural actors.** The local government can facilitate the exchange of good practices and knowledge acquired though the rural-urban partnership, promoting evaluation of the initiatives and sharing information.

Source: Adapted from OECD (2013), *Rural-Urban Partnership: An Integrated Approach to Economic Development*, OECD Publishing, Paris, http://dx.doi.org/10.1787/9789264204812-en.

People

Stakeholder engagement

Stakeholder engagement has historically mainly occurred at the sub-national and local levels as regions and cities operate at a more manageable scale for many forms of stakeholder engagement. They provide useful vehicles for translating national policy design and implementation at the local level, increasing national policy effectiveness. Also, regions and cities are responsible for much of the service delivery and public investment that determines economic growth and people's well-being.

Indeed, many policy experiments, such as living labs, smart cities, participatory budgeting or co-production of social service delivery, are carried out at the city or regional level. For example, many regions have established regional councils that combine elected officials, businesses, social partners and other relevant stakeholders, such as universities, which contribute to developing regional development strategies and oversee implementation. This is the case of the city of Grenoble, for example, where citizens play an important role in setting water tariffs and water-related investment priorities (Box 5.8).

Box 5.8. **Stakeholder engagement in setting tariffs and investment priorities in Grenoble**

In Grenoble, the water and sanitation service provider (Eau de Grenoble) engages with consumer associations to co-decide water prices. In 1996, a committee of water and sanitation users was created following a citizen initiative led by the NGO "Eau Secours" which had criticised abnormal tariff evolution following the privatisation of water provision in 1989. Today, an agreement between the committee and the current public service provider stipulates the roles, responsibilities and modalities of the joint activities related to information sharing as well as deliberations on water tariffs and water quality. Amongst others, the utility provides the venue for the four to six annual meetings, the budget for specific experts when needed, as well as other logistical expenses. The committee plays the role of advisor to the mayor on all measures concerning public water service provision, paying particular attention to issues related to the protection of the resource as well as fair prices for water. Every year, water managers and the committee of users discuss investments planned, the budget and the tariffs that would support needed infrastructure repairs in the network, while remaining affordable for all categories of users – domestic and industrial customers, social housing, business, etc.). The committee also has a seat at the managing board of the city's technical department for water, which manages the annexed water budget.

Source: OECD (2015d), *Stakeholder Engagement for Inclusive Water Governance*, OECD Studies on Water, OECD Publishing, Paris, http://dx.doi.org/10.1787/9789264231122-en.

It is increasingly recognised that enhanced engagement in local water-related decisions leads to active collaboration, innovative government-citizen partnerships and ultimately increased access to water and better cost-effective service delivery. Inclusive institutions at the local level can facilitate access to decision making; promote transparency, openness and engagement; avoid policy capture; and contribute to aligning policies with the needs to citizens, including social groups. There is growing recognition that services work better when designed and delivered in partnership with citizens, and that listening to stakeholders' insights can foster innovation in service delivery practices and better risk management (Box 5.9). In doing so, inclusive city administrations legitimise government actions and set a foundation for successful policy making and implementation, thus allowing a focus on medium- and long-term planning, an essential feature of effective water policy making.

Box 5.9. **Stakeholder engagement in cities**

Stakeholder engagement for tackling climate change

The Bologna Local Urban Environment Adaptation Plan for a Resilient City (BLUE AP) consists in setting up adaptation measures to meet climate change challenges, including specific measures for water management to cope with water scarcity and floods. The project started in 2012 and concluded in 2015. The City of Bologna planned and tested measures, which will be possibly replicated in other Italian cities. The project is co-ordinated by the municipality of Bologna and foreseen the technical collaboration of three partners: Ambiente Italia, ARPA Emilia Romagna e Kyoto Club.

Stakeholder engagement is an important part of the project. The first step consisted in the stakeholders mapping, which identified relevant stakeholders: public entities, private and public enterprises, university and research centres, specialised agencies, service providers, multi-utility, consortia, associations, consumers associations, environmental associations, and foundations. The engagement path consisted in one year of consultation, through several steps: a plenary meeting (2013) to present the documents elaborated to tackle climate change; thematic sessions with technical experts, to share practices, experiences and investigate on specific issues; in-depth sessions to evaluate the concrete implementation of received proposals; concluding plenary (2014), to report results to citizens and elaborate a strategic document for the Adaptation Plan. For each project, focus groups were set to elaborate technical sheets to be included in the Adaptation Plan. Overall 150 stakeholders have been involved, 70 project ideas have been presented and 6 pilot actions.

Stakeholder engagement to improve local services

In 2009, Nantes Métropole held a citizens workshop on the management of drinking water. Participants discussed key issues related to: i) information and training on principles of drinking water management; ii) research; and iii) decisions related to water. They subsequently issued an opinion during a public presentation to suggest 20 proposals: 11 refer to improving the operation of the Consultative Committee on Local Public Services (CCSPL), while 9 were centred on wider arrangements to facilitate citizen dialogue across the metropolitan area. The workshop showed that collective understanding and fruitful dialogues on how to improve and develop local public services can help to overcome the complexities and technicalities that typically characterise these services. The workshop reinforced the city's commitment to deepen the triangular relationship between the responsible authority, the operators, and users and citizens. By stimulating the Consultative Committee on Local Public Services, by initiating research on social tariffs, and by deepening the debate on how to articulate water resources and water services, the city intends to give full effect to this approach for better governance in a context of dynamic territorial development.

Box 5.9. **Stakeholder engagement in cities** *(continued)*

Paris' Water Observatory is an extra-municipal commission created in 2006 by the Mayor of Paris as a tool for collaboration and citizens' oversight regarding Paris' water policy. The Observatory is informed of all major deliberations on water management, on which it issues an opinion prior to their presentation at the City Council. These deliberations concern, for instance, the price and quality of public drinking water and sanitation services as well as the annual activity report of Eau de Paris (service provider). The Observatory is composed of four boards representing: i) elected officials at municipal level; ii) consumers; iii) local institutions from the health, urban planning, land use and housing sectors; and iv) academia. Beyond institutional and professional actors, any individuals or associations can join. The Observatory is also represented on the board of Eau de Paris and has voting power.

Partnerships with community groups in Singapore

The Active, Beautiful, Clean Waters (ABC Waters) Programme is a long-term initiative to transform Singapore's waterways and waterbodies beyond their traditional functions of drainage, flood control and water storage into scenic waterscapes and focal community points that offer a host of recreational options, hence inspiring a sense of collective ownership over water. To encourage the co-creation of ABC Waters projects, PUB engages the community from the early stages of project development to ensure that the sites are built based on what the community wants, keeps them updated about project progress, and works with them to make the sites more meaningful to the community. Over 100 potential projects have been identified for implementation in phases by 2030. A number of initiatives are carried out to engage various stakeholders. Through fostering advocacy and awareness of the water cause as well as adoption of water bodies, the engagement motivates the public community to conserve water and keep catchments and waterways clean so that they can enjoy the recreational opportunities it offers.

Source: Input provided by the City of Bologna and Singapore through the OECD Survey on Water Governance for Future Cities (2014) and Comune di Bologna (n.d.), "BLUE AP Project, Strategia di adattamento ai cambiamenti climatici della Città di Bologna", available at: www.blueap.eu/site/wp-content/uploads/2013/09/BLUEAP_Strategia_adattamento_locale.pdf; PUB (2013), "Our Water, our future", available at: http://www.pub.gov.sg/mpublications/OurWaterOurFuture/Documents/OurWaterOurFuture_2015.pdf; Observatoire parisien de l'eau website (n.d.), www.observatoireparisiendeleau.fr, (accessed November 2015); Eau de Paris website (n.d), www.eaudeparis.fr, (accessed October 2015).

Stakeholder engagement is key for co-ordinating various actors and interests in cities. It is herein defined as the process through which individuals, groups, and organisations have the opportunity to take part in the decision making that will affect them, or in which they have an interest (OECD, 2015d). Stakeholder engagement can bring together urban planners, water service providers, regulators, advisors and civil society to develop dynamic integrated approaches. As a governance instrument, stakeholder engagement can help build trust and ownership, secure the willingness to pay for water services, raise awareness on current and future water challenges ensure the accountability of city managers and service providers to end users and citizens, manage conflicts on water allocation, ensure the political acceptability of different ownership models, and set convergent objectives across policy areas.

City departments interact with various authorities when it comes to managing water. Their main counterparts are service providers (for 81% of respondents), followed by regional governments (69%) and local governments (60%). The interaction with central governments is less frequent (it takes place "sometimes" for 33%, and "never or rarely" for 13% of respondents). There is also a rather low interaction with irrigators and their associations (never, for 40% of cities surveyed), civil society and business/industry ("sometimes" 40% and 35%, respectively) (Figure 5.7).

Figure 5.7. **Frequency of interactions between cities and stakeholders**

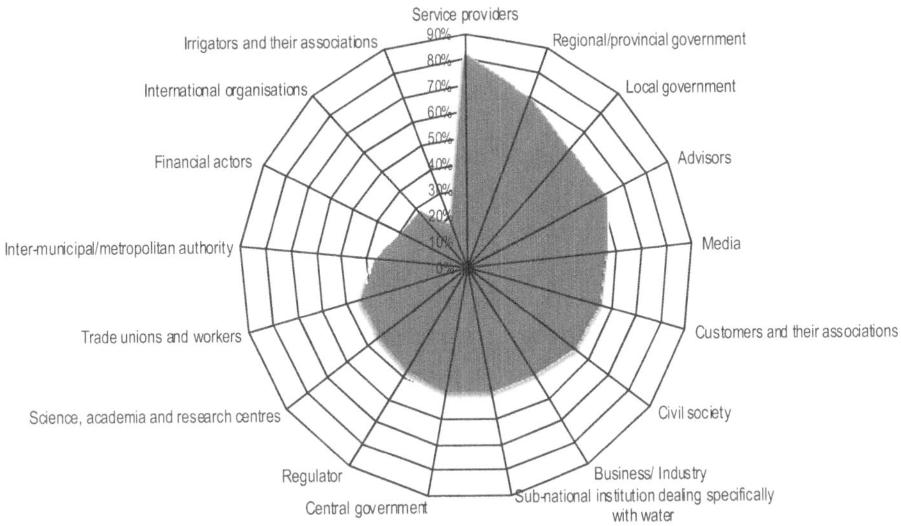

Note: Results based on the responses provided as "always, very frequently", "often" out of a scale from "always or very frequently", "often", "sometimes" to "very rarely" or 'never". The sample includes 48 respondents.
Source: OECD Survey on Water Governance for Future Cities (2014).

Obstacles to engaging stakeholders

There are several obstacles hindering the effective contribution of stakeholder engagement to decision making related to water management in cities (Figure 5.8). Almost half of surveyed cities pointed out the complexity of issue at hand, the resistance to change, in addition to the lack of funding to support stakeholder engagement (44%). Other challenges are the difficulty to reach certain types of stakeholders and lack of time (42%). Interestingly, lacking or ineffective stakeholder engagement is also due to the stakeholders themselves and not necessarily to the lack of conditions for engaging them (38% of surveyed cities pointed out the consultation fatigue and misaligned objective of stakeholders); and the geographical distance from the decision-making core does not appear as a major obstacle, nor the decision makers fear of losing their influence and power.

Figure 5.8. **Obstacles to stakeholder engagement**

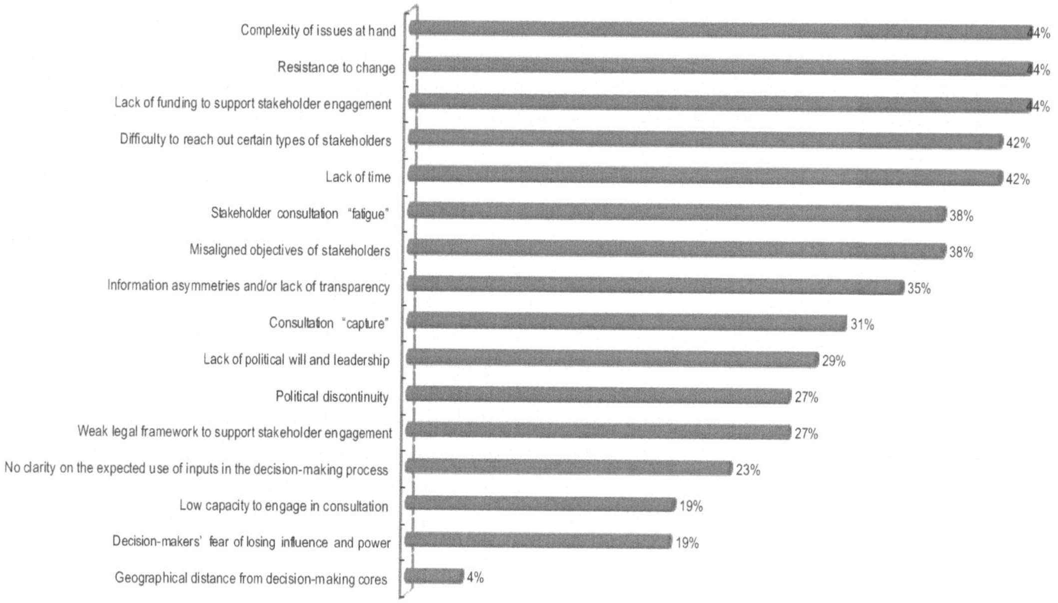

Note: Results based on a sample of 48 respondents who indicated the obstacles being "critical" and "important".

Source: OECD Survey on Water Governance for Future Cities (2014).

Effective communication and education are also crucial for stakeholder engagement, as without awareness there is no consensus. They are a necessary tool to let public (users) and other stakeholders understand non-technical and technical issues. For instance, greater understanding of the urban water cycle and related costs would raise awareness on the economic value of water. Being aware of complications and costs for having safe water from the tap and securing sanitation systems is a first step needed for stakeholder engagement, which can be carried out by education and awareness campaigns for different audiences. This would also make the decision process more inclusive and build consensus on sensitive issues, such as that of the price of water. Despite wide recognition that inclusive approaches contribute to good water governance, establishing the conditions for lasting changes can be challenging. Water policy has traditionally been carried out from the top down: from the perspective of bureaucrats, officials, politicians rather than from the perspective of citizens, businesses and consumers. Even if the importance of inclusive institutions is gaining traction and has become an appealing concept to understand citizens' needs and improve trust in government, there are still important blockages that prevent change from taking place beyond procedural practices.

There is also a need to dispel the myths around stakeholder engagement. Even with the best intentions and framework conditions in place, stakeholder engagement may fail to achieve an inclusive process and may even entail capture if not properly handled. Ultimately, there is an overall lack of systematic evaluation, hence a lack of evidence on the performance of institutions to support inclusive water-related decision making. Concretely, more evidence on the impacts of stakeholder engagement and practical guidance on successful practices taking into consideration contexts and place-based specific dimensions are much needed.

Mechanisms to engage with stakeholders

Cities can choose from a continuum of engagement mechanisms. Modalities vary from basic communication of information, which is the lightest form of engagement, to full co-production, co-delivery and co-evaluation, which involves a balanced share of powers among stakeholders. Each of these approaches has different goals and may have different impacts, and they thus require different tools for implementation. The modalities for stakeholder engagement vary in nature. The OECD Survey on Stakeholder Engagement in the Water Sector (OECD, 2015d) identified 24 mechanisms that can be categorised into two types: formal mechanisms that have institutional and legal ground, and informal mechanisms that are not institutionalised but can be implemented for a large variety of issues and at the discretion of the convener of the engagement process. These different mechanisms have both strengths and weaknesses (Tables 5.2 and 5.3). Moreover, engagement modalities vary in terms of the amount of time they take, the number of stakeholders they involve and the amount of resources they require. Similarly, different tools may apply best to different steps of the policy cycle (i.e. design, implementation or evaluation) or to different categories of actors.

The steady integration of new technologies, including social media and mobile technology, is giving rise to new forms of public engagement. This new digital environment offers opportunities for more collaborative and participatory relationships that allow relevant stakeholders to actively shape political priorities, collaborate in the design of public services, and participate in their delivery to provide more coherent and integrated solutions to complex challenges. This is particularly relevant in the light of research that suggests that citizens want to be heard and feed into the early stages of decision making rather than claiming power in the final stages (Ker Rault et al., 2013; Gharesifard and Wehn, 2015).

Digitally enabled participation and production of services is changing people's expectations about their relationships with local governments. The OECD Recommendation on Digital Government Strategies (OECD, 2014d) highlights the need for new public governance approaches to support a shift from governments anticipating citizens' and business's needs (citizen-centric approaches) to citizens and businesses determining their own needs and addressing them in partnership with governments (citizen-driven approaches).

Among the sample of cities surveyed, technology plays an important role in engaging stakeholders, with 83% of respondents using web-based communication technologies (online platforms, e mail, social media, websites and apps) on a regular basis to engage with stakeholders (Figure 5.9). Information and communications technologies (ICTs) can foster stakeholder engagement: for instance, citizen-based observations of water levels play an increasing role in water management. Moreover, metrics, graphics, and GIS (Geographic Information System) (used by water utilities to display information) allow customers to view their water bills in near real time, be informed about the level of water quality, and/or to alert them of a potential surge of usage suggesting a leak (e.g. as in the case of the new information system of Eau de Marseille).

Table 5.2. **Assessment of formal mechanisms for stakeholder engagement**

Mechanisms	Strengths	Weaknesses
Citizen committee	– Allow the involvement and input of a range of stakeholders – Allow development of consensus (where achievable) or directions for action on complex issues – Provide opportunities for exploring alternative strategies – Stakeholders gain an understanding of other perspectives leading toward an agreed, integrated outcome.	– Participant selection is a major consideration. The range of interests must be broad enough to represent all those affected, and those with relevant interests and skills. Organisers must be aware of potential conflicts. – The general public may not embrace committee recommendations. – Members may not achieve consensus (although consensus may not be the goal). – Can be time- and labour-intensive if the issue is significant.
Consensus conference	– Empower stakeholders to develop an informed understanding and make some contribution to the development of policy and projects – Demonstrate a plurality of views on issues – Bridge the gap between experts and less-knowledgeable stakeholders – Can develop new knowledge	– High costs for set up and recruitment of participants – The selection process can be difficult. Mapping stakeholders is critical to predetermine the relevant groups.
Innovative contracts and partnerships	– Foster co-ordination and co-operation across stakeholders and potentially levels of governments – Help manage interdependencies – Can solve institutional weaknesses	– Unclear objectives and allocation of tasks among partners or signatories may lead to inefficiency – Can be time- and labour-intensive
Interest-pay-say principle	– Stakeholders engaged are often highly motivated to contribute in return for their financial contribution/ investment.	– Appointing (minority) representatives is sometimes perceived as "less democratic"
Polls/survey	– Provide traceable data – Can serve an educational purpose – When properly constructed using good sampling techniques, can reach a broad, representative public or targeted group.	– Poorly constructed surveys produce poor results – Careful sampling is needed to make sure representative samples are taken.
Referendum	– Provide a representative view of a population's opinion on a specific issue – In the context of voter apathy and disenchantment with traditional forms of democracy, direct democracy can help to re-engage voters with policy matters.	– Voters do not always have the capacity or information to make informed decisions about the issue at stake, and instead may make ill-informed decisions based on partial knowledge or on the basis of unrelated factors.
River basin organisations/ councils	– Wide public and stakeholder participation in decision making – Local empowerment	– Deliberative decision making may be dominated by a small group of stakeholders (farmers, industries, etc.) – Legal frameworks for setting up river basin organisations do not always provide for the engagement of stakeholders in the decision making – Need for substantial financial and human resources to be sustainable
Stakeholder democracies	– Stakeholder groups have a direct say in all decisions taken by the assembly, including financing issues.	– Risk of low participation rates in elections in context of awareness gap regarding water-related issues – Stakeholder groups represented in the assembly have to be well organised.
Water associations	– Common understanding across members of the issues at stake – Often, high-level of expertise from experienced practitioners – Legitimacy of the association to represent the views of its members when engaging with public authorities and other stakeholders	– Can be perceived as single-minded when they solely focus on pushing forward the agenda of a singular group of stakeholders (e.g. association of irrigators) – May not encompass a wider membership that includes other players likely to be impacted by their activities

Source: OECD (2015d), *Stakeholder Engagement for Inclusive Water Governance*, OECD Studies on Water, OECD Publishing, Paris, http://dx.doi.org/10.1787/9789264231122-en.

Table 5.3. **Assessment of informal mechanisms for stakeholder engagement**

Mechanisms	Strengths	Weaknesses
Expert panel	– Useful when an issue is complex and contentious – Useful where conflict exists to provide opinions which may have more credibility, and hence may assist in resolving the conflict	– Expertise in relevant and complementary areas may be needed to produce a credible expert opinion. – Skilled moderator is often required
Focus group	– Produce ideas that would not emerge from surveys/questionnaires, because the focus group allows opportunity for a wider range of comments	– Such small groups may not be representative of the community response to an issue – Require careful selection to be a representative sample
Information hotline	– Offer an inexpensive and simple device for publicity, information and public input – Can serve as a link between the citizens and the municipality's government	– Must be adequately advertised to be successful – Can be time-consuming to manage and update on a regular basis
Meetings/workshops/fora	– Allow the involvement and input of a wide range of stakeholders – Disseminate detailed information and decisions across stakeholders – Can build ownership and credibility for the outcomes – Contribute to better communication among the stakeholders involved	– Can be time- and labour-intensive
Stakeholder mapping	– Provide detailed stakeholder analysis (motivations and interests, interactions, scale of intervention)	– Can be time-consuming – Can be based on subjective data and may vary according to the person/place
Traditional media	– Can disseminate information quickly to a large number of people	– Difficult to retract should any changes occur – The size of media releases limit the amount of real content that can be incorporated.
Web-based technology	– Capable of reaching very large numbers of stakeholders with very large amounts of information – Allow participants to discuss an issue at their convenience (regardless of location or time) – Anonymity of online processes can encourage open discussion	– Many people still cannot access the Internet. – Information overload and poor design can prevent people from finding what they need.

Source: OECD (2015d), *Stakeholder Engagement for Inclusive Water Governance*, OECD Studies on Water, OECD Publishing, Paris, http://dx.doi.org/10.1787/9789264231122-en.

Figure 5.9. **Existing mechanisms to engage stakeholders in surveyed cities**

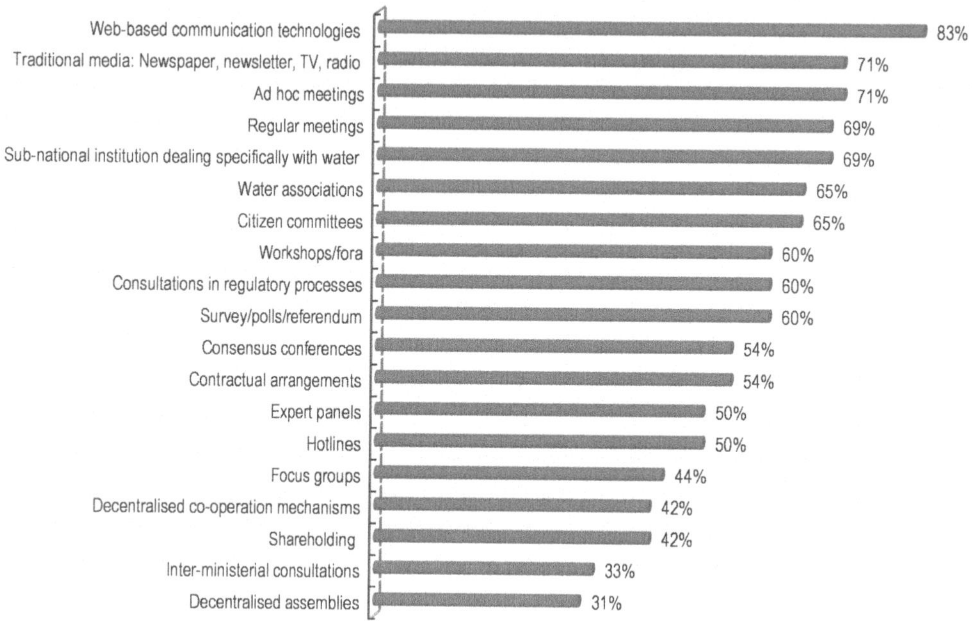

Note: Results based on a sample of 48 respondents who responded "yes" to the options provided.

Source: OECD Survey on Water Governance for Future Cities (2014).

Despite the diversity of stakeholder engagement modalities and tools, a number of public governance levers are common and essential enablers of effective, meaningful engagement. They include: i) knowing the stakeholders; ii) having information that is accessible, relevant and usable; and iii) using technologies that facilitate timely, cost-effective interactions and feedback. The effectiveness of engagement mechanisms also depends on capacities and resources, including know-how and funding.

Governments at all levels have a critical role to play in establishing the enabling environment for results-oriented, effective and impactful stakeholder engagement. Knowing that engagement processes cannot be easily replicated from one context to another, OECD (2015d) suggests six basic principles to set up the proper framework conditions for results-oriented stakeholder engagement, and to catalyse efforts for making good practices more visible (Box 5.10). These principles can be used with a Checklist for Public Action to support the implementation, consisting in priority questions and a set of indicators to help governments and stakeholders monitor the effectiveness of engagement processes and identify areas of improvement.

Box 5.10. **Principles for effective stakeholder engagement in water decision making**

- **Map all stakeholders who have a stake in the outcome or that are likely to be affected, as well as their responsibility, core motivations and interactions.** Stakeholder mappings should be done in relation to a specific issue and be updated on a regular basis. Such mapping should pay attention to newcomers, players outside the water sector and traditionally under-represented groups. This is critical to ensure that all stakeholders are identified and properly involved throughout the policy/project cycle. Finding the right balance between inclusiveness and empowerment of stakeholders is also important. Engagement processes (and related mechanisms) need to accommodate the needs of stakeholders with varying levels of interests and resources to ensure inclusivity and accessibility. Risks related to consultation capture from over-represented categories to the detriment of unheard voices, as well as risks of prejudice to a particular category of stakeholders, deserve careful consideration. Equity between present and future generations in a perspective of sustainability should be promoted. Thus, disaggregated data on gender, age economic status and the level of impact of proposed policies and measures is crucial.
- **Define the ultimate line of decision making, the objectives of stakeholder engagement and the expected use of inputs.** Clarifying the goals and reasons for engagement is key to building mutual understanding and trust of how stakeholders may be involved in the process, and for informed stakeholders to provide quality contributions in line with expectations. Objectives of stakeholder engagement can contribute to the formulation of river basin plans at the watershed level, service delivery, awareness-raising (e.g. on water costs, risks, future trends), auditing, risk mapping, as well as performance monitoring. Similarly, the authority responsible for taking decisions, and its willingness to take stakeholders' ideas on board in doing so, should be clearly identified to enhance confidence in the value of the process. Transparency and accountability in how the engagement process is designed and implemented (e.g. stakeholder mapping methods, use of stakeholders' inputs) is crucial to improve credibility and legitimacy, and build trust among the stakeholders involved. Diligent work is necessary to ensure that the engagement process is fair and equitable and to reliably engage stakeholders.
- **Allocate proper financial and human resources and share needed information for result-oriented stakeholder engagement**. Improving the overall contribution to substantive discussions and decision making requires access to timely and understandable information (be it cultural, scientific, traditional, etc.), technical expertise, experience sharing and funding in the right format and sufficiently on time (planning) to realistically and effectively participate. Supporting two-way information sharing through consistent and appropriate communication channels, including web-based technologies when feasible, is key. Ensuring the financial affordability of the engagement process is also important, to ensure the effective engagement of all those that have a stake; convey accurate, trusted and accessible information to diverse sectors; foster opinion forming within and across stakeholder groups; and build support to the process. The interpretation and application of these resources and information require competences and capability development at all levels to enable sustainable stakeholder engagement (e.g. skills, social learning).
- **Regularly assess the process and outcomes of stakeholder engagement to learn, adjust and improve accordingly.** Such evaluation and monitoring can use fact-based and perception-based tools and indicators, and be carried out by targets, promoters and/or third parties. Results should be disclosed to increase accountability, provide insight on the success of the engagement process in reaching its intended objectives and learn from experience to improve practice in the future. Evaluation should not be limited to *ex ante* and *ex post* assessment but remain an ongoing process throughout the policy/project cycle. Stakeholder engagement can yield benefits in terms of resilience, sustainability, cohesion, acceptability, capacity and efficiency. But it can also delay decision making and implementation, and generate different types of (monetary and non-monetary) material, process, reputational and social costs. Assessing the costs and benefits of engagement processes can help to ensure that all interests, including under-represented stakeholders, are respected regarding the distribution of impacts, compensation and benefits. Mitigation measures are needed to reduce costs and set the right incentives while managing the dual short-term/long-term temporality.

Box 5.10. **Principles for effective stakeholder engagement in water decision making** *(continued)*

- **Embed engagement processes in clear legal and policy frameworks, organisational structures/principles and responsible authorities.** There is no water governance without governance at large. Similarly, there can be no effective stakeholder engagement without proper incentives for bottom-up and inclusive policy making. A clear set of rules, platforms and vehicles for doing so is critical to move from reactive to proactive and systematic stakeholder engagement in the water sector. But institutionalisation increases the risk of engagement "fatigue" and/or "capture" from over-represented categories to the detriment of unheard voices. It should be flexible to take into consideration place-based needs and changing circumstances while fostering a change in the "mindset", daily practices, professional skills and culture of decision making. Provisions for stakeholder engagement should be aligned coherently and holistically across the water chain and policy domains related to water.
- **Customise the type and level of engagement to the needs and keep the process flexible to changing circumstances.** Stakeholder engagement tools and mechanisms work differently across places, times, objectives and stages of the policy/project cycle. They should be tailored to each context (geographic, socio-economic, cultural), type of stakeholder concerned, policy goal targeted and place-based needs to accommodate varying levels of interest and resources from stakeholders and consider other options as needs arise. Water governance systems are complex and in flux, where change is dynamic and often unpredictable. Engagement processes, therefore, need to enable multiple stakeholders to respond and adapt to uncertainty and should remain flexible to manage risks and resilient to adapt to changing environments. Lessons can be learnt from failure in engagement approaches in terms of management of complexity and how to trigger fundamental change.

Source: OECD (2015d), *Stakeholder Engagement for Inclusive Water Governance*, OECD Studies on Water, OECD Publishing, Paris, http://dx.doi.org/10.1787/9789264231122-en.

Forward-looking strategies for urban water governance

Drawing from the result of the Survey on Water Governance for Future Cities, three priorities stand out clearly as the most common across respondents to cope with future challenges: ensuring value for money (higher quality at lower cost) (63%), building, operating and/or maintaining water infrastructure (60%) and raising awareness on water availability, risks, quality and costs (56%) (Figure 5.10). It is however noteworthy to mention that surveyed cities above 5 million inhabitants identified technical and non-technical innovation as a top priority as well.

Figure 5.10. **Top priorities to cope with future water-related challenges**

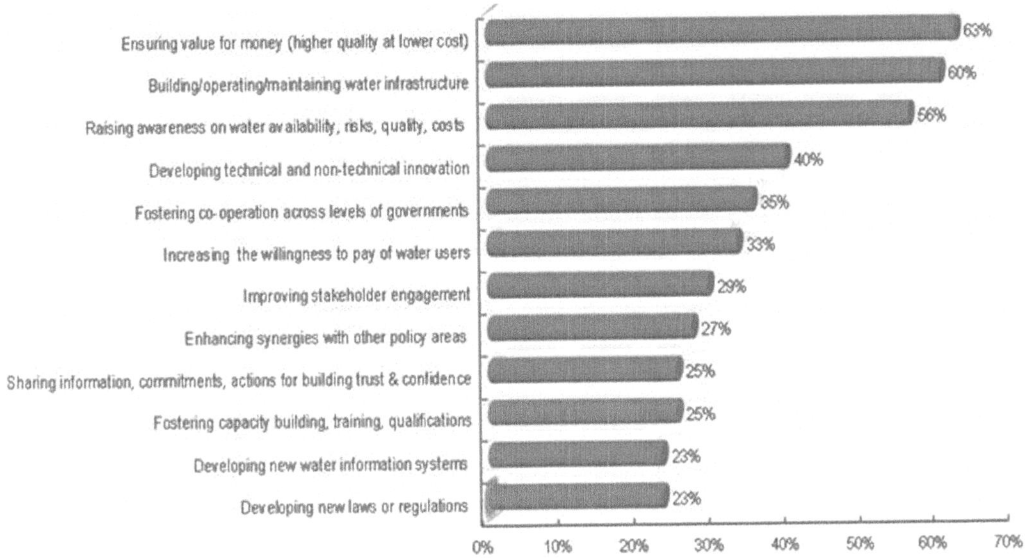

Note: Results based on a sample of 48 respondents who indicated the actions having "top priority".

Source: OECD Survey on Water Governance for Future Cities (2014).

The OECD Principles on Water Governance (OECD, 2015c) "promote the adoption and implementation of innovative water governance practices across responsible authorities, levels of government and relevant stakeholders" (Principle 8). This is a means for cities to favour the implementation of experimentation and pilot-testing on water governance (e.g. through contracts across levels of government); promoting social learning to facilitate dialogue and consensus-building (e.g. through networking platforms, social media, ICTs, etc.); finding innovative ways to co-operate, to pool resources and capacity, to build synergies across sectors and search for efficiency gains (e.g. through metropolitan governance, inter-municipal collaboration, urban-rural partnerships, and performance-based contracts); and fostering science-policy interface to bridge the divide between scientific findings and water governance practices. Examples of innovation in the water sector are provided in Box 5.11: technical innovations are often accompanied by governance measures to raise awareness and build consensus.

Box 5.11. **Innovation in the water sector**

Recycled water in Singapore

In 2003, the Public Utilities Board (PUB), Singapore's national water agency, introduced NEWater as one of Singapore's Four National Taps (which also include local catchment water, imported water and desalinated water). It is high-grade reclaimed water produced from treated used water that has undergone stringent purification and treatment process using advanced dual-membrane (microfiltration and reverse osmosis) and ultraviolet technologies. It has passed over 130 000 scientific tests and exceeds the drinking water standards set by the World Health Organization and the US Environmental Protection Agency. NEWater is used primarily for non-potable industrial purposes at wafer fabrication parks, industrial estates and commercial buildings. During dry months, NEWater is used to top up the reservoirs and blended with raw water before undergoing treatment at the waterworks before being supplied for the drinking water supply.

Prior to the development of NEWater, Singapore had to rely heavily on local catchments and imported water from Johor in Malaysia as its key water sources. However, these two traditional sources are weather-dependent. While reclaiming used water is not a new concept, what is significant for Singapore is the wide-scale implementation and widespread public acceptance of NEWater for indirect potable use. This is part of an overall strategy to raise awareness of the population, stressing a new approach to water management by communicating to the public the need to look at water as a renewable resource that could be used over and over again. The price of NEWater is cheaper than that of potable water and this has encouraged many industries to switch to NEWater. Strict enforcement of used water discharge also plays an important role in ensuring that water reclamation plants are able to function as designed, which then supply part of the treated effluent to the NEWater plants. Water reclamation technology is relevant to other water-scarce regions. From an energy perspective, it is about one quarter of what desalination would require. It is from this perspective that NEWater holds tremendous promise for developing cities.

The Total Water Management (TWM) Strategy in Hong Kong, China

The Total Water Management (TWM) Strategy was promulgated in 2008 to better prepare Hong Kong, China for uncertainties and to enhance Hong Kong's role as a good partner to other cities in the Pearl River Delta in promoting sustainable use of water.

The TWM Strategy has proven accomplishments towards water security and reliability in Hong Kong, China. The TWM Strategy put emphasis on containing growth of water demand through promotion of water conservation and strengthening water supply and demand management. Key measures include enhancing public education on water conservation, promoting use of water-saving devices, enhancing water leakage control, extending use of seawater for toilet flushing, strengthening protection of water resources, actively considering water reclamation, and developing the option of seawater desalination. A comprehensive review of the TWM Strategy is being carried out to ensure sustainable use of water resources and timely introduction of new initiatives to strengthen resilience and preparedness against uncertainties and challenges.

Source: Inputs provided by the surveyed cites. More information available at The Government of Hong Kong Special Administrative Region (2008), Total Water Management in Hong Kong, available at: http://www.wsd.gov.hk/filemanager/en/share/pdf/TWM.pdf and PUB (n.d), Singapore's National Water Agency website, http://www.pub.gov.sg/water/newater/Pages/default.aspx, (accessed November 2015).

Ways forward for stronger local-national frameworks for managing water in cities

There is no unique solution to cope with water-related challenges but a range of responses according to contexts, cities' human, financial and technical capacity and the priority given to water in their broader agenda. Cities are not isolated and need to implement win-win approaches with surrounding areas to manage interdependences; they have to enhance the buy-in, trust and confidence by engaging stakeholders and to favour policy complementarities and avoid silos to reap the benefits of integrated urban water management.

Building on the OECD Principles on Water Governance, a number of policy conclusions can be drawn to foster stronger urban-national policy frameworks that can help to better manage too much, too little and too polluted water in cities and their hinterland. These actions can be carried out at: i) local level primarily by cities; ii) in co-operation between cities and their hinterland; and iii) in co-operation with upper levels of government (Box 5.12).

Box 5.12. **The OECD Principles on Water Governance**

The OECD Principles on Water Governance frame 12 "must-haves" for governments to reap the economic, social and environmental benefits of good water governance (OECD, 2015a). They apply to all levels of government, regardless of water management functions, water uses, and ownership models, and are grouped into three clusters:

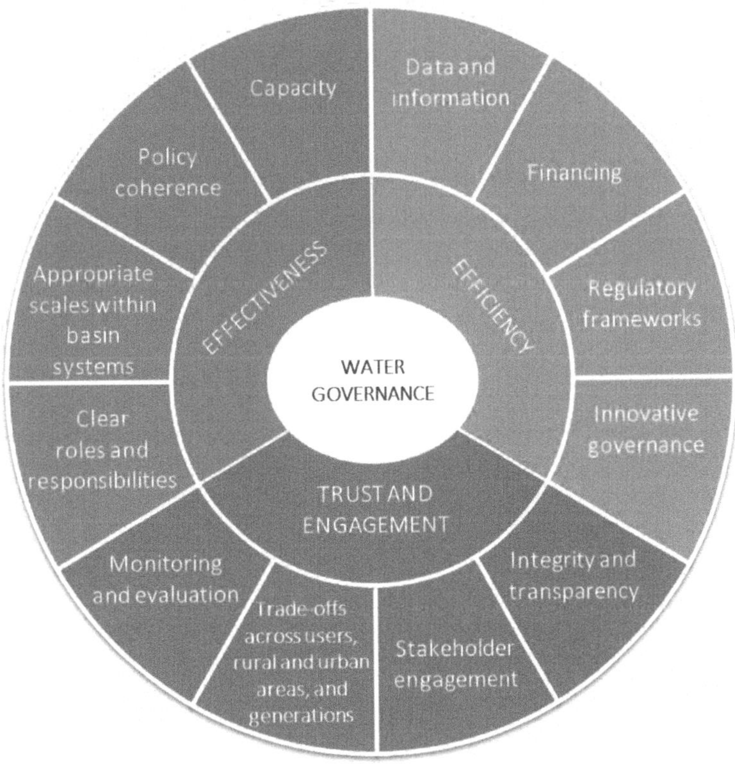

- **Effectiveness of water governance** relates to the contribution of governance to define clear sustainable water policy goals and targets at different levels of government, to implement those policy goals, and to meet expected objectives or targets.

Box 5.12. **The OECD Principles on Water Governance** *(continued)*

- **Efficiency of water governance** relates to the contribution of governance to maximise the benefits of sustainable water management and welfare at the least cost to society.
- **Trust and Engagement in water governance** relate to the contribution of governance to building public confidence and ensuring inclusiveness of stakeholders through democratic legitimacy and fairness for society at large.

Source: OECD (2015a), "OECD Principles on Water Governance", OECD, Paris, available at: www.oecd.org/gov/regional-policy/OECD-Principles-on-Water-Governance-brochure.pdf.

A prominent role for cities

Making the most of policy complementarities. Governance mechanisms can favour policy complementarities between water and other sectors, such as agriculture, energy, environment and spatial planning. They can reconcile administrative, functional and hydrological logics, overcome territorial mismatch; efficiently allocate resources (e.g. for alternative use of water as in the energy sector); and increase capacity (e.g. by combining management of multiple sectors – waste, water, energy).

Making the most of stakeholder engagement. As water-related risks are increasingly worrying policy makers, financial resources are scarce and trust in government in many countries is at low levels, stakeholder engagement can be build trust and ownership, secure the willingness to pay for water services, raise awareness on current and future water risks, manage conflicts on water allocation and set convergent objectives across policy areas. Stakeholders all have a role to play in urban water management, alongside policy makers.

Making the most of multi-level approaches to water governance. Water management takes place at several scales, from basin to local levels, depending on the service required (protection against floods or droughts, water supply, sanitation, drainage, etc.) and it is scattered across a number of actors. Co-ordination is crucial to minimise overlaps, duplications and to identify grey areas, which can hinder the effectiveness of water policies. Metropolitan governance and rural-urban partnerships are two powerful instruments to enhance co-ordination at subnational level.

Making the most of the city-to-city learning process. Cities can learn lessons and best practices through city-to-city knowledge networks. There are several platforms to share innovative successful practices and experiences in the water sector; including through networks or umbrella organisations such as C40, 100+ Resilient Cities, the Sustainable Cities Network, ICLEI, H2O, or the City Blueprint. In the Danube region the IAWD (International Association of Water Supply Companies in the Danube River Catchment Area) is also a good example for knowledge exchange. The Green Cities programme of the OECD[7] also built a "knowledge sharing platform" to share best practices from the OECD member countries with Thailand, Philippines and Viet Nam. This can inspire similar initiatives on in other areas, such as water governance.

A shared responsibility between cities and their hinterland

Making the best of rural-urban co-operation. Rural areas can benefit from a positive two-way dialogue with urban areas, but this requires mutual trust and a clear understanding of the long-term benefits of the interaction. Strengthening *capacities*, especially in rural areas, is key. Improving the *information system* and transparency is also needed as in many rural areas; the status and flows of groundwater remain largely unknown. Rural areas are also often disadvantaged by weak communications infrastructure and in some cases digital divide. Managing trade-offs related to *fairness and equity* in access to resources and services is also key to ensure that general and specific interests (i.e. farmers, companies, nature reserves, etc.) are heard and the voice of rural citizens is brought to the table to reflect rural poor' interests.

A need for joint undertaking between cities and upper levels of government

Decentralisation and increasing autonomy of cities are making sub-national governments key actors of urban policies and closer to citizens' needs. Political decentralisation can lead to increased political accountability and transparency, and fiscal decentralisation can improve public spending effectiveness based on the idea that sub-national governments have better information on local spending needs and preferences and are better positioned to deliver public goods. Political and fiscal decentralisation while strengthening the role of sub-national governments still require co-operation with national government. Decentralisation can come at a cost, in terms of funding or capacity gaps at the subnational level; in practice it is not about autonomy, but partnerships.

In the case of urban water governance, a national/ sub-national co-operation is needed for:

- **Building capacities**: The national government can help empower the sub-national one if competences are new and the capacity of sub-national government is limited. When dealing with new competences in water, as a result of early stages of decentralisation, sub-national governments may lack expertise. Central governments may "empower" sub-national ones through capacity-building tools that favour learning and dialogue (e.g. contracts across levels of governments).

- **Developing innovative practices**: In the case of mature decentralisation, there can be a mutual learning, by the central and the sub-national government in developing shared innovative practices throughout to the various stages of the water policy cycle, from planning to evaluation. This can help improve the quality of the policy choices (i.e. better tailored to sub-national specificities as a result of enhanced strategic planning efforts or more rigorous ex-ante appraisal).

- **Policy alignment**: When managing water, sub-national authorities and central governments are "mutually dependent". Policy alignment is needed because policy at national level holds impacts on lower levels of government. Drawing from other more general issues at local level, some conclusions can be applied to the water sector in terms of policy coherence. Special attention needs to be paid to legislative frameworks for cities: national state/provincial legislations typically define cities' responsibilities, powers and, crucially, revenue sources, but attention to the basic legislative framework for cities is often overlooked. There is a need to focus on potential, not only on problems: often national governments intervene in policy domains that affect cities, in response to specific problems such as infrastructure bottlenecks. The broader needs of cities are thus overlooked, as well as possible synergies with other policies. Last, co-ordination is key: leadership from higher levels of government is often required to favour co-operation among municipalities, which is much needed, in particular in the case of metropolitan areas to do better with less.

- **Financial support**: When decentralisation results in insufficient resources available to lower levels of government to carry out their tasks, sub-national governments may financially depend on central governments. Among surveyed cities, for example, only a few have an autonomous budget for investment in the water sector. At the same time, central and sub-national governments can be co-funders of public spending in the water sector in regions.

- **Improving performance**: While devolving responsibilities to sub-national governments, central governments can reinforce the control on standards of water services or on performance of service delivery, and transparency over who pays for what and how pricing policy is carried out.

Notes

1. See: EPA Website (n.d.), Water Infrastructure and Resiliency Finance Center, http://www.epa.gov/waterfinancecenter, (accessed December 2015).

2. Decree 5 July 2012 n. 391/Ric.

3. See Eau de Paris (2014), Dossier de Presse - Quand eau et agriculture biologique s'allient pour un développement local durable 30 bonnes raisons de passer à l'action - 4 juillet 2014, available at: www.eaudeparis.fr/uploads/tx_edpevents/DPFnabEDPvdef.pdf, (accessed November 2015).

4. See Scottish Water (2013), The Metropolitan Glasgow Strategic Drainage Partnership, Briefing Note 10 - Winter 2012/13, available at: http://www.scottishwater.co.uk/assets/domestic/files/investment%20and%20communities/mgsdp/mgsdpjanfebnewsletter10.pdf.

5. There are 71 Optimal Territorial Areas (ATO) in Italy, but further mergers are likely to occur as a consequence of the territorial reform concerning the recent creation of metropolitan bodies (*Città Metropolitane*). This will be the case of Milan: by 2016, the ATO governing body of the Metropolitan Area will perform its duties also on the territory of the City of Milan, which has been regulated so far by a distinguished ATO governing body (Zelioli, 2015).

6. See Vodakva website, The Water Management Association of the West Bohemia Region, www.vodakva.cz/en/about-company/structure/shareholders.html, (accessed November 2015).

7. See OECD (n.d.), Green City Programme, OECD, Paris, available at: http://www.oecd.org/gov/regional-policy/Programme_Description_OECD_Urban_Green_Growth_in_Dynamic_Asia.pdf

Bibliography

Alexandratos, N. and J. Bruinsma (2012), "World agriculture towards 2030/2050, the 2012 revision", *ESA Working Paper,* No. 12-03, June 2012, Agricultural Development Economics Division Food and Agriculture Organization of the United Nations.

Ahrend, R., C. Gamper and A. Schumann (2014), "The OECD Metropolitan Governance Survey: A quantitative description of governance structures in large urban agglomerations", *OECD Regional Development Working Papers*, No. 2014/04, OECD Publishing, Paris, http://dx.doi.org/10.1787/5jz43zldh08p-en.

Àrea metropolitana de Barcelona website (n.d.), www.amb.cat, (accessed October 2015).

Autorità d'ambito Torinese website (n.d.), www.ato3torinese.it/loader.html?organigramma/conferenza.html, (accessed December 2015).

Charbit, C. and O. Romano (forthcoming), "Contracts across levels of government for regional development: An international review", paper submitted to the Chilean Sub-secretariat for Regional Development (SUBDERE), June 2015.

City of Montreal (2007), "Le Réseau Bleu – Montréal vers une culture de l'eau", Reference Framework, July.

Comune di Bologna (n.d.), BLUE AP Project, Strategia di adattamento ai cambiamenti climatici della Città di Bologna, available at: www.blueap.eu/site/wp-content/uploads/2013/09/BLUEAP_Strategia_adattamento_locale.pdf.

Depuratore Nosedo website (n.d.), www.depuratorenosedo.eu, (accessed October 2015).

Eau de Paris (2014), Dossier de Presse - Quand eau et agriculture biologique s'allient pour un développement local durable 30 bonnes raisons de passer à l'action - 4 juillet 2014, www.eaudeparis.fr/uploads/tx_edpevents/DPFnabEDPvdef.pdf, (accessed November 2015).

Eau de Paris website (n.d.), www.eaudeparis.fr, (accessed September 2015).

EPA Website, Water Infrastructure and Resiliency Finance Center, http://www.epa.gov/waterfinancecenter, (accessed December 2015).

FAO (2011), *Global Food Losses and Food Waste – Extent, Causes and Prevention*, FAO, Rome.

Gharesifard, M. and U. Wehn (2015), "Participation in citizen science: Drivers and barriers for sharing personally-collected weather data via web-platforms", paper presented at the 8th International Conference on ICT, Society and Human Beings (ICT2015), 21-23 July, Las Palmas de Gran Canaria, Spain.

IEA (2012), *World Energy Outlook 2012*, OECD Publishing, Paris, http://dx.doi.org/10.1787/weo-2012-en.

IEA (2009), "Energy efficiency requirements in building codes, energy efficiency policies for new buildings", available at: www.iea.org/publications/freepublications/publication/ Building_Codes.pdf.

Ker Rault, P. et al. (2013), "Readiness and willingness of the public to participate in integrated water management: Some insights from the Levant", *Water Policy*, Vol. 15/2, pp.101–120.

Marest, P. et al. (2006), "Sustainable water management, the choices of Nantes Metropole". In Urban Public Services: the triptych organising authority, operators, citizen-users, Local powers n°71 III/ December 2006

Observatoire parisien de l'eau website (n.d.), www.observatoireparisiendeleau.fr, (accessed November 2015).

OECD (2015a), "OECD Principles on Water Governance", brochure, OECD, Paris, available at: www.oecd.org/gov/regional-policy/OECD-Principles-on-Water-Governance-brochure.pdf.

OECD (2015b), *Governing the City*, OECD Publishing, Paris, http://dx.doi.org/10.1787/ 9789264226500-en.

OECD (2015c), *Water and Cities. Ensuring Sustainable Futures*, OECD Publishing, Paris, http://dx.doi.org/10.1787/9789264230149-en.

OECD (2015d), *Stakeholder Engagement for Inclusive Water Governance*, OECD Studies on Water, OECD Publishing, Paris, http://dx.doi.org/10.1787/9789264231122-en.

OECD (2014a), *Compact City Policies: Korea: Towards Sustainable and Inclusive Growth*, OECD Green Growth Studies, OECD Publishing, Paris, http://dx.doi.org/10.1787/9789264225503-en.

OECD (2014b), *Water Governance in the Netherlands: Fit for the Future?,* OECD Studies on Water, OECD Publishing, Paris, http://dx.doi.org/10.1787/9789264102637-en.

OECD (2014c), *OECD Regional Outlook 2014: Regions and Cities: Where Policies and People Meet*, OECD Publishing, Paris, http://dx.doi.org/10.1787/9789264201415 en.

OECD (2014d), "Recommendation on Digital Government Strategies", www.oecd.org/gov/public-innovation/recommendation-on-digital-government-strategies.htm, (accessed October 2015).

OECD (2013), *Rural-Urban Partnerships: An Integrated Approach to Economic Development*, OECD Publishing, Paris, http://dx.doi.org/10.1787/9789264204812-en.

OECD (2012a), *OECD Environmental Outlook to 2050: The Consequences of Inaction*, OECD Publishing, Paris, http://dx.doi.org/10.1787/9789264122246-en.

OECD (2012b), *Water Quality and Agriculture: Meeting the Policy Challenge*, OECD Studies on Water, OECD Publishing, Paris, http://dx.doi.org/10.1787/9789264168060-en.

OECD (2011), *Water Governance in OECD Countries: A Multi-level Approach*, OECD Studies on Water, OECD Publishing, Paris, http://dx.doi.org/10.1787/9789264119284-en.

OECD (2009), *Governing Regional Development Policy: The Use of Performance Indicators*, OECD Publishing, Paris, http://dx.doi.org/10.1787/9789264056299-en.

OECD (2007), *Linking Regions and Central Governments: Contracts for Regional Development*, OECD Publishing, Paris, http://dx.doi.org/10.1787/9789264008755-en.

OECD (1999), "Intergovernmental partnerships at the local level in Spain: Mancomunidades and consortia in a comparative perspective", OECD, Paris, available at: http://www.municipal-cooperation.org/images/4/49/Paper_Intergovernmental_Partnerships_at_the_Local_Level_Spain_1999.pdf

OECD (n.d.), "Green City Programme", brochure, OECD, Paris, available at: http://www.oecd.org/gov/regional-policy/Programme_Description_OECD_Urban_Green_Growth_in_Dynamic_Asia.pdf

PUB (2013), Our Water, our future, available at: http://www.pub.gov.sg/mpublications/OurWaterOurFuture/Documents/OurWaterOurFuture_2015.pdf.

PUB (n.d.), Singapore's National Water Agency website, http://www.pub.gov.sg/water/newater/Pages/default.aspx, (accessed November 2015).

Saussier S. and M. Klein (2014), "Local Public Enterprises: A Taxonomy", *EPPP DP* No. 2 014-12, available at: www.chaire-eppp.org/files_chaire/a_taxonomy_of_lpe.pdf.

Scottish Water (2013), The Metropolitan Glasgow Strategic Drainage Partnership, Briefing Note 10 - Winter 2012/13, available at: http://www.scottishwater.co.uk/assets/domestic/files/investment%20and%20communities/mgsdp/mgsdpjanfebnewsletter10.pdf.

The Government of Hong Kong Special Administrative Region (2008), Total Water Management in Hong Kong, available at: http://www.wsd.gov.hk/filemanager/en/share/pdf/TWM.pdf.

Vodakva website (n.d.), The Water Management Association of the West Bohemia Region, www.vodakva.cz/en/about-company/structure/shareholders.html, (accessed November 2015).

Watershed Agricultural Council website (n.d.), www.nycwatershed.org, (accessed September 2015).

Zelioli A. (2015), Sfide future per una nuova governance. La Regolazione locale di secondo livello del Servizio Idrico Integrato della Città di Milano verso la Città Metropolitana, Servizi a rete, Novembre-Dicembre 2015, available at: http://www.atocittadimilano.it/docs/news/sar%2012.2015.pdf

ANNEX A

Respondents to the OECD Survey on Water Governance for Future Cities

City	Respondent	Organisation	Website
Acapulco	Augustin Ceballos Contreras	CAPAMA, Comisión de agua potable y alcantarillado de municipio de Acapulco, Depto. de Evaluación de Proyectos de Saneamiento y Potabilización	www.capama.gob.mx
Amsterdam	Freek Brink and Geurt Rombach	Waternet	www.waternet.nl
Athens	Olga Ntri	EYDAP S.A, Department of Development of New Activities	www.eydap.gr
Barcelona	Esther Suárez Albi	Area Metropolitana de Barcelona	www.amb.cat
Belo Horizonte	Ricardo de Miranda Aroeira	Municipality of Belo Horizonte, DUDECAP/PBH	www.pbh.gov.br
Bologna	Giovanni Fini	Città di Bologna, Dipartimento Riqualificazione Urbana_Settore Ambiente ed Energia	www.comune.bolgna.it
Budapest	Gabor Kisvardai	Hungarian Energy and Public Utility Regulatory Authority, Secretariat of the Vice-President for Public Utilities	www.mekh.hu/en
Calgary	Maria Samson and Maggie Choi	City Of Calgary, Water Resources	www.calgary.ca/
Chihuahua	Carlos Daniel Alonso Guzman (2015)/Maurilio Ochoa(2014) and Jorge Macias/Manuel Altes/Carmen Julia Navarro	Junta Municipal de Agua y Saneamiento de Chihuahua	www.jmaschih.gob.mx
Cologne	Jutta Lenz and Matthias Schmitt	Stadtentwesserungsbetriebe Koln/RheinEnergie AG	www.steb-koeln.de; www.rheinenergie.com
Copenhagen	Søren Povlsen	HOFOR Vand KÃ¸benhavn A/S	www.hofor.dk
Culiacan	Cesar Arechiga Torres	JAPAC, Junta Municipal de agua potable y alcantarillado de Culiacan,Dpto de estadisticas e informacion	www.Japac.gob.mx
Daegu	Park Hee Sun	Daegu Metropolitan City, Waterworks Headquarters	
Edinburgh	Jon Rathjen, Barry Greig and Catherine McKenna	Scottish Government Water Industry Team	www.scotland.gov.uk
Glasgow	Jon Rathjen, Barry Greig and Catherine McKenna	Scottish Government Water Industry Team	www.scotland.gov.uk
Grenoble	Patricia Bajard (2014) Jacques Tcheng (2015)	ISERE - Grenoble Régie de l'eau potable SPL eau de Grenoble Alpes	www.eauxdegrenobleapes.fr
Hermosillo	David Contreras (2014)/ Renato Ulloa Valdes (2015)	Agua de Hermosillo	www.aguadehermosillo.gob.mx
Hong Kong,China	Richard Wm Leung Alex Kk Chan Gabriel Kf Pang	Drainage Services Department Environmental Protection Department Water Supplies Department	www.dsd.gov.hk www.epd.gov.hk www.wsd.gov.hk
Kitakyushu City	Hiromon Mariko (2014) / Koba Yukinori (2015)	Kitakyushu City Water Supply and Sewerage Bureau Division	
Krakow	Wacław Skubida (2014) / Malgorzata Starnowska (2015)	Urząd Miasta Krakowa – Municipality of Krakow Wydział Gospodarki Komunalnej – Municipal Services Department	gk.umk@um.krakow.pl
Lisbon	Jose Sardinha	EPAL, Empresa Portuguesa de Águas Livres	www.epal.pt
Liverpool	Steven Holcroft	Liverpool City Council	www.liverpool.gov.uk
Malaga	Jose Luis Rodriguez Lopez	EMASA, Empresa Municipal De Aguas De Málaga	www.emasa.es

City	Respondent	Organisation	Website
Marseille	Jean-Marc Mertz	Marseille Provence Métropole, Services urbains de proximité	www.marseille-provence.com
Mexico City	Ramon Aguirre Diaz	SACMEX, Sistema de aguas de la ciudad de Mexico	www.sacmex.df.gob.mx
Milan	Andrea Zelioli	Ufficio d'Ambito della Città di Milano	www.atocittadimilano.it
Montreal	Chantal Morissette	Ville de Montréal, Service de l'eau	http://ville.montreal.qc.ca/eaudemontreal
Nantes	Philippe Marest	Nantes Métropole, Département environnement et services urbains	www.metropole.nantes.net
Napoli	Maurizio Giugni	Assessorato alle Infrastrutture, Lavori Pubblici e Mobilità	
New York City	Hannah Thonet David Lipsky (point of contact)	New York City Department of Environmental Protection	www.nyc.gov/dep Contact: dlipsky@dep.nyc.gov
Okayama	Hideki Nagata	Okayama City, Sewerage and River Bureau, Sewerage Management Planning Division	
Oslo	Arnhild Krogh and Frode Hult	City of Oslo, Oslo Water and Sewerage Works	www.vann-og-avlopsetaten.oslo.kommune.no
Paris	Odile Nieuwyaer (2014)/ Anne du Plessis (2015)	Ville de Paris, Chargée des relations avec les usagers de l'eau	www.paris.fr
Phoenix	Kathryn Sorensen	City of Phoenix, Water Services Department	www.phoenix.gov
Prague	Eva Bartonickova	Prague Institute for Planning and Development (IPR Prague)	www.iprpraha.cz
Queretaro	José Eduardo Martin Pérez Salinas (2014)/ Salvador Espinoza Medina (2015)	Comisión estatal de aguas de Queretaro, Planeación Estratégica	www.ceaqueretaro.gob.mx
Rio de Janeiro	João Luiz Reis da Silva	Fundação Instituto das Águas do Município do Rio de Janeiro (Rio-Águas)	www.rio.rj.gov.br/web/rio-aguas
Rome	Roberto Zocchi and Francesco Bosco	Acea S.p.A., Azienda Comunale Energia e Ambiente	www.acea.it
San Luis Potosi	Ezequiel Duran de Anda	INTERAPAS, Unidad de Proyectos	www.interapas.gob.mx
Singapore	Michelle Ooi (2014)/ Xueyi Liao (2015) and Ming Hwee Lee, Liao Xueyi, Tiing Liang Moh, Tay Kai Yun	PUB, Singapore's National Water Agency	www.pub.gov.sg
Stockholm	Lars Lindblom Juha Salonsaari Malin Parmander	Stockholm Water AB City of Stockholm City of Stockholm	www.stockholm.se
Suzhou	Xia Jian, Gu Ming, Huang Hanyi, Gao Lei, Zhang Yna, Wu Hua, Che Meiqin	Suzhou Water Resources Bureau	
Toluca	Juan Mario Dominguez Alonso (2014)/ José Maya Ambrosio (2015)	Organismo agua y saneamiento del municipio de Toluca	www.ayst.gob.mx
Turin	Paolo Romano and Armando Quazzo	SMAT S.p.A, Società Metropolitana Acque Torino	www.smatorino.it
Tuxtla	Jose Alfredo Araujo Esquinca (2014)/ Carlos Flores (2015)	SMAPA, Sistema Municipal de Agua Potable y Alcantarillado de Tuxtla Gutierrez	www.smapa.gob.mx
Veracruz	Juan Manuel Ruz Barros	SAS, Sistema de Saneamiento de Veracruz/ Unidad de medición, evaluación y control	www.nueva.sas.gob.mx
Zaragoza	Javier Celma/Carmen Cebrián	Ayuntamiento de Zaragoza, Agencia de Medio Ambiente y Sostenibilidad	www.zaragoza.es
Zibo	Xu Fawen	Water Resources Management Office of Zibo	www.zbszy.gov.cn/WebPortal/index.aspx

ORGANISATION FOR ECONOMIC CO-OPERATION AND DEVELOPMENT

The OECD is a unique forum where governments work together to address the economic, social and environmental challenges of globalisation. The OECD is also at the forefront of efforts to understand and to help governments respond to new developments and concerns, such as corporate governance, the information economy and the challenges of an ageing population. The Organisation provides a setting where governments can compare policy experiences, seek answers to common problems, identify good practice and work to co-ordinate domestic and international policies.

The OECD member countries are: Australia, Austria, Belgium, Canada, Chile, the Czech Republic, Denmark, Estonia, Finland, France, Germany, Greece, Hungary, Iceland, Ireland, Israel, Italy, Japan, Korea, Luxembourg, Mexico, the Netherlands, New Zealand, Norway, Poland, Portugal, the Slovak Republic, Slovenia, Spain, Sweden, Switzerland, Turkey, the United Kingdom and the United States. The European Union takes part in the work of the OECD.

OECD Publishing disseminates widely the results of the Organisation's statistics gathering and research on economic, social and environmental issues, as well as the conventions, guidelines and standards agreed by its members.

OECD PUBLISHING, 2, rue André-Pascal, 75775 PARIS CEDEX 16
(42 2016 02 1 P) ISBN 978-92-64-25108-3 – 2016

IWA Publishing's authorised EU representative for General Product Safety Regulations is Diane D'Arras, 15 rue Duret, 75116 Paris, France, e-mail: safety@iwap.co.uk.

Printed and bound by CPI Group (UK) Ltd, Croydon, CR0 4YY
12/05/2026
02108886-0012